SIDREA Series in Accounting and Business Administration

This is the official book series of SIDREA - the Italian Society of Accounting and Business Administration. This book series is provided with a wide Scientific Committee composed of Academics by SIDREA. It publishes contributions (monographs, edited volumes and proceedings) as a result of the double blind review process by the SIDREA's thematic research groups, operating at the national and international levels. Particularly, the series aims to disseminate specialized findings on several topics – classical and cutting-edge alike – that are currently being discussed by the accounting and business administration communities. The series authors are respected researchers and professors in the fields of business valuation; governance and internal control; financial accounting; public accounting; management control; gender; turnaround predictive models; non-financial disclosure; intellectual capital, smart technologies, and digitalization; and university governance and performance measurement. This book series is indexed in Scopus.

Andrea Caccialanza

Sustainability and Accountability of Supply Chains

Research Avenues on the Interplay Between Regulations, Standards and Emerging Practices

Springer

Andrea Caccialanza
University of Bologna
Bologna, Italy

ISSN 2662-9879 ISSN 2662-9887 (electronic)
SIDREA Series in Accounting and Business Administration
ISBN 978-3-032-16878-8 ISBN 978-3-032-16879-5 (eBook)
https://doi.org/10.1007/978-3-032-16879-5

This Springer imprint is published by the registered company Springer Nature Switzerland AG
The registered company address is: Gewerbestrasse 11, 6330 Cham, Switzerland

Introduction

As countries strive to address grand challenges stemming from climate change, policymakers have turned their scrutiny on the sustainability and accountability of supply chains (Aureli et al., 2020; Gray et al., 1996). As a consequence, companies are under increasing pressure to incorporate assessments of their non-economic performance into strategic planning and thereby orient their decision-making around the sustainability orientation of managerial practices (Sarkis et al., 2016). In the current globalization context, supply chains are a key competitive driver in various industries (Altman & Bastian, 2021; Brinkmann, 2023). At the same time, scholarly and policy debates now converge on the idea that sustainability is produced and distributed across multi-tier value chains, not just within organizational boundaries—thus requiring a meso-level lens that makes inter-firm impacts measurable, comparable, and actionable across tiers (Beske & Seuring, 2014; Altman & Bastian, 2021). This shift motivates the present study's focus on supply chain accountability, where managerial practices, data architectures, and institutional standards interact to shape outcomes beyond the focal firm. In fact, the constant pursuit of technological innovation is accelerating competitive pressures, pushing firms to update their products and processes—which often requires leaning on supply chain partners. At the same time, the public's increasing awareness of supply chains' ethical practices and social commitments has prompted regulatory bodies and organizations to enact measures geared toward greater accountability (Akkartal & Mızrak, 2023; Remko, 2020).

Against that background, regulators have moved to update international frameworks for assessing corporate performance. These increasingly stringent requirements seek to push companies to implement new metrics that can satisfy the information needs of multiple stakeholder categories. In this context, the European Financial Reporting Advisory Group (EFRAG) recently introduced the European Single Reporting Standard (ESRS), which focuses on enhancing accountability within supply chains across the European Union (EU) (EFRAG, 2022–2023 a-b). The proposed ESRS aims to heighten transparency and accountability in supply chain management—particularly concerning environmental, social, and governance (ESG) factors (Parrondo, 2024)—in the hopes of mitigating the effects of climate change (Martin-Ortega, 2022; Parrondo, 2024). Moreover, the ESRS seeks to provide a standardized framework for reporting on various aspects of supply chain accountability, including—but not limited to—labor practices, environmental impact, human rights, and corporate governance (EFRAG, 2022).

By aligning reporting practices across EU member states, the ESRS seeks to facilitate comparability and consistency in supply chain disclosures. This would render businesses' climate change impacts more explicit (Bassetti et al., 2024; Beske & Seuring, 2014; Khalid et al., 2015; Khan et al., 2021) and hopefully enhance the reliability and credibility of their disclosures (Khalid & Seuring, 2019). This standardized approach can provide valuable information to stakeholders (e.g., investors, consumers, and regulatory authorities), while also promoting fair competition and responsible business practices within the European market. Ideally, those changes would have ripple effects on non-EU firms that are part of European supply chains (Balon, 2020; Dai et al., 2021). In this way, the ESRS reflects the EU's commitment to advancing sustainable development goals (SDGs), promoting corporate social responsibility (CSR) initiatives (Balon, 2020; Blome et al., 2014) and establishing corporate accountability benchmarks that encompass the double materiality approach (Dodsworth et al., 2023; Setia et al., 2024). Early evidence nonetheless indicates that value-chain analysis is the least mature area of implementation: many undertakings begin at tier-1 and expand gradually, with the binding constraint on comparability and assurance-readiness appearing to be data integration across tiers (EFRAG, 2024).

In short, the new ESRS represents a substantial attempt to enhance accountability in supply chains within the European context. This framework underscores the importance of regulatory initiatives in promoting sustainable economic growth and advancing societal objectives in the modern sustainable supply chain management field (Khan et al., 2021; Mardani et al., 2020). Given the significance of this development, both scholars and practitioners are called upon to deeply explore its potential impact (Gauzès & Board, 2021; Martin-Ortega, 2022; Parrondo, 2024). However, this has exposed a notable gap: Much of the prior literature and many legacy frameworks remain firm-centric, under-representing how sustainability performance is generated across multi-tier networks and how evidence governance beyond tier-1 conditions the credibility of disclosures (Beske & Seuring, 2014; Khan et al., 2021). This monograph therefore asks:

How does the ESRS shift accountability from the single firm (micro) to the supply chain (meso), and through which institutional levers and data mechanisms can that accountability be made proportionate, comparable and decision-useful?

To address this question, the study adopts an integrative literature review across SSCM, accounting and institutional regulation, combined with a documentary analysis of the CSRD/ESRS corpus, the Omnibus proposals, and EFRAG outputs. It leverages EFRAG's initial observed practices (Q2–2024) to gauge implementation readiness and the [Draft] Amended ESRS 2 (2025) to clarify expectations on IRO-first processes and the cautions against boilerplate (EFRAG, 2024; 2025). To aid that process, this monograph offers a broad overview of supply chain dynamics: first defining the role of accounting and accountability in sustainability performance, and then focusing on how different institutions contribute to improving supply chains. The institutional role is no longer limited solely to defining the (minimum) disclosure requirements or identifying the entities obliged to publish their sustainability performance. Instead, it now plays an active role, for example, in setting standards, identifying key topics of focus at both the supply chain and firm levels.

Building on these inputs, Chap. 5 develops a framework—see Fig. 19—that integrates (i) the sustainability of the value chain, (ii) the accountability of the value chain, and (iii) the institutional level (EFRAG/ESRS) into a single meso-level lens. The framework operationalizes how chain-specific evidence travels across tiers and how proportionality (e.g., value-chain cap; VSME) stabilizes information requests without lowering ambition (EFRAG, 2024a; 2024b; 2025). The extensive literature review herein follows a twofold objective: to provide (i) a concise yet comprehensive overview of the current debate regarding the integration of sustainability and accountability principles in supply chains; and (ii) a structured research agenda on the topic in order to outline future research avenues. This contribution especially focuses on the European context, highlighting the inbound and outbound externalities resulting from the implementation of the ESRS guidelines (Kazancoglu et al., 2021; Tsai et al., 2021). The conceptual framework that emerges from the monograph's Chap. 5 is the result of the combination of various debated topics. Concretely, the review (i) maps the debate on integrating sustainability and accountability at meso-level, (ii) specifies an agenda for chain-aware evaluation and disclosure under the ESRS, and (iii) anticipates three results discussed later: that double materiality and value-chain scope formalize meso-level accountability; that the Omnibus re-sequences and right-sizes implementation while retaining the core; and that evidence governance beyond tier-1 is decisive for comparability and assurance-readiness (EFRAG, 2024; 2024a; 2024b; 2025).

The remainder of the book is organized into five subsections that reflect the main topics discussed in the research agenda:

1. The foundations of sustainable supply chain management and procurement;
2. The role of accounting and accountability principles;
3. The role of institutions and norms;
4. The literature review focused on the evaluation and disclosure of sustainability performance, specifically from the micro-level and meso-level perspectives;
5. Research avenues to foster the sustainability and accountability of supply chains.

This roadmap is designed to link the research aim to the evidence base: foundational sustainable supply chain management and accounting principles set the stage for the institutional analysis; the micro-to-meso transition grounds the ESRS and "Omnibus Package" discussion; and the research avenues translate the framework into testable implications for sectors and supply networks with different tier structures. The target audience of the volume naturally includes scholars in the supply chain management and accounting domain, who may leverage the framework to address the proposed research questions or further the debate. However, the results may also hold value for practitioners and business consultants who are currently implementing the ESRS and need to analyze sustainability performance. For practitioners, the analysis clarifies how to improve the supply chain accountability while preserving the value-chain cap as risk-based prioritization criteria, and how to use European reporting standards (i.e., the VSME or the ESRS) as a baseline to stabilize supplier requests, while investing in traceability and supplier enablement to improve assurance-ready, chain-level indicators from the first reporting cycles onward (EFRAG, 2024a; 2024b).

References

Akkartal, G. R., & Mızrak, F. (2023). Defining reverse and green supplier selection criteria with a strategic management approach. In Economic Development and the Environmental Ecosystem: The Role of Energy Policy in Economic Growth (pp. 205–218). Cham: Springer Nature Switzerland.

Altman, S. A., & Bastian, C. R. (2021). The state of globalization in 2021. *Harvard Business Review*, 3, 1–9.

Aureli, S., Del Baldo, M., Lombardi, R., & Nappo, F. (2020). Nonfinancial reporting regulation and challenges in sustainability disclosure and corporate governance practices. *Business Strategy and the Environment*, 29(6), 2392–2403.

Bassetti, T., Dal Maso, L., & Pieroni, V. (2024). Firms' borrowing costs and neighbors' flood risk. *Small Business Economics*, 1–17.

Beske, P., & Seuring, S. (2014). Putting sustainability into supply chain management. *Supply Chain Management: An International Journal*, 19(3), 322–331.

Blome, C., Hollos, D., & Paulraj, A. (2014). Green procurement and green supplier development: antecedents and effects on supplier performance. *International Journal of Production Research*, 52(1), 32–49.

Brinkmann, R. (Ed.). (2023). The Palgrave Handbook of Global Sustainability. Springer Nature.

Dai, J., Xie, L., & Chu, Z. (2021). Developing sustainable supply chain management: The interplay of institutional pressures and sustainability capabilities. *Sustainable Production and Consumption*, 28, 254–268.

Dodsworth, C., Bourne, E., Schlich, B. L., & Kooroshy, J. (2023). Financial, Double, or Dynamic? Theories of ESG Materiality and Practitioner Approaches. *Journal of Impact & ESG Investing*, 4(2).

EFRAG (2022). [Draft] EFRAG IG 3 List of ESRS Data Points (Last update: 23/12/2022). Retrieved from: https://efrag.sharefile.com/share/view/s1a12c193b86d406e90b1bcd7b6bb8f6f/fo37c90b-9d9b-4432-a76b-27760cfcc01b

EFRAG (2023a). [Draft] EFRAG IG 1: Materiality assessment implementation guidance. Retrieved from: https://www.efrag.org/Assets/Download?assetUrl=%2Fsites%2Fwebpublishing%2FSiteAssets%2FDraft%2520EFRAG%2520IG%25201%2520MAIG%2520231222.pdf

EFRAG (2023b). [Draft] EFRAG IG 2: Value chain implementation guidance. Retrieved from: https://www.efrag.org/Assets/Download?assetUrl=%2Fsites%2Fwebpublishing%2FSiteAssets%2FDraft%2520EFRAG%2520IG%25202%2520VCIG%2520231222.pdf

EFRAG (2024). Implementation of ESRS: Initial Observed Practices from Selected Companies. Available at: https://www.efrag.org/sites/default/files/media/document/2024-07/EFRAG_ESRS%20initial%20observed%20practices%20Q2%202024%20final%20version.pdf

EFRAG (2024a). EFRAG IG 1 Materiality Assessment (Last update: May 2024). Retrieved from: https://www.efrag.org/lab3?AspxAutoDetectCookieSupport=1#subtitle6

EFRAG (2024b). ESRS for VSMEs (voluntary standard). Retrieved from: https://www.efrag.org/lab3?AspxAutoDetectCookieSupport=1#subtitle6

EFRAG. (2025). [Draft] Amended ESRS 2—Exposure Draft—July 2025. Brussels: European Financial Reporting Advisory Group. Available at: https://www.efrag.org/sites/default/files/media/document/2025-07/Amended_ESRS_Exposure_Draft_July_2025_ESRS_2.pdf.

Gauzès, J. P., & Board, E. (2021). Potential Need of Changes to the Governance and Funding of EFRAG. European Reporting Lab. Retrieved from: https://leasenews.it/storage/app/uploads/public/604/643/d36/604643d36bcf0117093624.pdf

Gray, R., Owen, D., & Adams, C. (1996). Accounting and accountability: Changes and challenges in corporate social and environmental reporting. London: Prentice-Hall.

Kazancoglu, I., Sagnak, M., Kumar Mangla, S., & Kazancoglu, Y. (2021). Circular economy and the policy: A framework for improving the corporate environmental management in supply chains. *Business Strategy and the Environment*, 30(1), 590–608.

Khalid, R. U., & Seuring, S. (2019). Analyzing base-of-the-pyramid research from a (sustainable) supply chain perspective. *Journal of Business Ethics*, 155, 663–686.

Khalid, R. U., Seuring, S., Beske, P., Land, A., Yawar, S. A., & Wagner, R. (2015). Putting sustainable supply chain management into base of the pyramid research. *Supply Chain Management: An International Journal*, 20(6), 681–696.

Khan, S. A. R., Yu, Z., Golpira, H., Sharif, A., & Mardani, A. (2021). A state-of-the-art review and meta-analysis on sustainable supply chain management: Future research directions. *Journal of Cleaner Production*, 278, 123,357.

Mardani, A., Kannan, D., Hooker, R. E., Ozkul, S., Alrasheedi, M., & Tirkolaee, E. B. (2020). Evaluation of green and sustainable supply chain management using structural equation modelling: A systematic review of the state of the art literature and recommendations for future research. *Journal of Cleaner Production*, 249, 119,383.

Martin-Ortega, O. (2022). Transparency and human rights in global supply chains: from corporate-led disclosure to a right to know. In Research Handbook on Global Governance, Business and Human Rights (pp. 100–120). Edward Elgar Publishing.

Parrondo, L. (2024). EFRAG roadmap for new developments in ERS reporting. In Research Handbook on Financial Accounting (pp. 138–155). Edward Elgar Publishing.

Remko, V. H. (2020). Research opportunities for a more resilient post-COVID-19 supply chain–closing the gap between research findings and industry practice. *International Journal of Operations & Production Management*, 40(4), 341–355.

Sarkis, J., Hervani, A. A., & Helms, M. (2016). Applying Economic Non-Market Valuation for Sustainable Supply Chain Performance Measurement and Evaluation. In Sustainable Value Chain Management (pp. 197–219). Routledge.

Setia, N., Abhayawansa, S., Joshi, M., & Wasantha Pathiranage, N. (2024). Shifting perspectives: unveiling the dual nature of sustainability materiality in integrated reports. *Meditari Accountancy Research*, 32(4), 1291–1323.

Tsai, F. M., Bui, T. D., Tseng, M. L., Ali, M. H., Lim, M. K., & Chiu, A. S. (2021). Sustainable supply chain management trends in world regions: A data-driven analysis. *Resources, Conservation and Recycling*, 167, 105,421.

Acknowledgment

This study was funded by the European Union—Next Generation EU, in the framework of the "GRINS—Growing Resilient, INclusive and Sustainable project" (PNRR - M4C2 - I1.3 - PE00000018—CUP J33C22002910001). The views and opinions expressed are solely those of the authors and do not necessarily reflect those of the European Union, nor can the European Union be held responsible for them.

Competing Interests The author has no competing interests to declare that are relevant to the content of this manuscript.

Contents

List of Figures

Chapter 1
Fundamentals of Supply Chain Literature

1.1 Definitions of Supply Chains

This subsection discusses the key features of supply chains that underline the development of the discipline and informs the current debate on how to introduce sustainability practices into supply chain management.

In their seminal work, La Londe and Masters (1994) proposed that a "supply chain is a set of firms that pass materials forward. Normally, several independent firms are involved in manufacturing a product and placing it in the hands of the end user in a supply chain. Raw material and component producers, product assemblers, wholesalers, retailer merchants, and transportation companies are all members of a supply chain" (Mentzer et al., 2001, p. 3). Mentzer et al. (2001) attempted to link this definition to the historical evolution of the term, thus providing a review of concepts related to processes and stemming from product/service analyses. However, their work exclusively targeted the manufacturing context, specifically the process of physically moving materials along the supply chain (from raw material suppliers to final consumers). This strong association with logistics harkens back to a militaristic view of the supply chain, which focuses on optimizing the costs and resources required to transfer supplies to a given place by a given time. Several definitions support this prevalent approach: Just as troops need to receive supplies on time, the modern production context requires that manufacturing sites receive the correct quantities with the appropriate delivery methods. As a result, firms tend to emphasize the resilience and reliability of the process itself rather than its nonfinancial economic impact (Craighead et al., 2007; Pettit et al., 2019; Wiedmer et al., 2021).

Around the turn of the century, scholars and practitioners were looking for ways to enhance logistics efficiency, which naturally entailed more attention to supply chains (Pienaar, 2009). This rationale responded to a call for enhancing production processes at a time when efficiency improvements were beginning to receive more

A. Caccialanza, *Sustainability and Accountability of Supply Chains*, SIDREA Series in Accounting and Business Administration,
https://doi.org/10.1007/978-3-032-16879-5_1

attention (Christopher, 1999; La Londe & Masters, 1994; Oliver & Webber, 2012). Beamon (1998) answered this call by offering a more structured definition of supply chains: "as an integrated process wherein a number of various business entities (i.e., suppliers, manufacturers, distributors, and retailers) work together in an effort to: (1) acquire raw materials, (2) convert these raw materials into specified final products, and (3) deliver these final products to retailers. This chain is traditionally characterized by a forward flow of materials and a backward flow of information" (p. 281). Beamon (1998, p. 282) consolidated this definition into a framework (Fig. 1.1) that summarizes the logistics view of the supply chain.

Beamon's framework usefully rationalized the raw material flow in supply chains but scholars quickly realized the need to extend the model to information such as product-related data. Beamon's (1998) classic supply chain logistics representations were foundational for this discipline, but they remain largely aspatial: They abstracted away from the geographic dispersion and role differentiation that now define distributed production networks globally. Moreover, modern supply chains are marked by increasing complexity and a higher number of components in the manufacturing context. According to Gereffi et al. (2005, p. 98), "On the upstream end, component and equipment suppliers can wield a great deal of power (through their market dominance in key components and technologies), setting parameters that most other value chain actors must adjust to. On the downstream end of the chain, highly knowledgeable users determine the attributes and innovative trajectory of the products and services that global value chains churn out." Therefore, these approaches are no longer sufficient to offer substantial guidance on how (sustainability) data should move across tiers. Early network research began to surface structural complexity and geographically scattered multitier relationships (Choi & Hong, 2002), and sustainability reviews flagged the need to integrate environmental, social, and economic criteria across extended supplier bases, yet the informational requirements for doing so remained underdeveloped. Researchers have primarily explored the internal dynamics of flows (both material and informational) with the goal of improving productivity (Patnayakuni et al., 2006) and identifying production bottlenecks during the

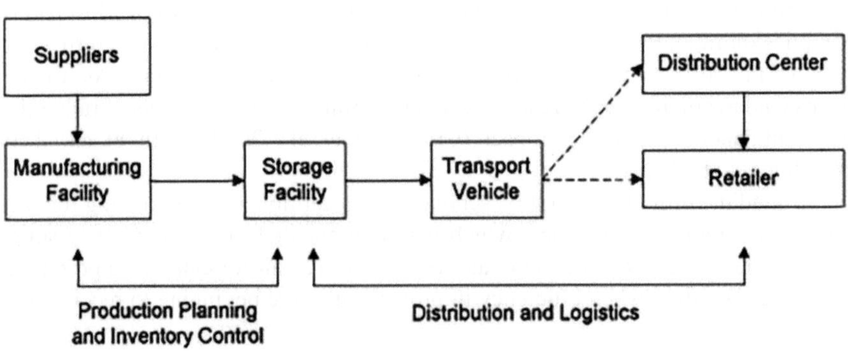

Fig. 1.1 Beamon (1998) supply chain (logistic) process. Source: Beamon (1998, p. 282)

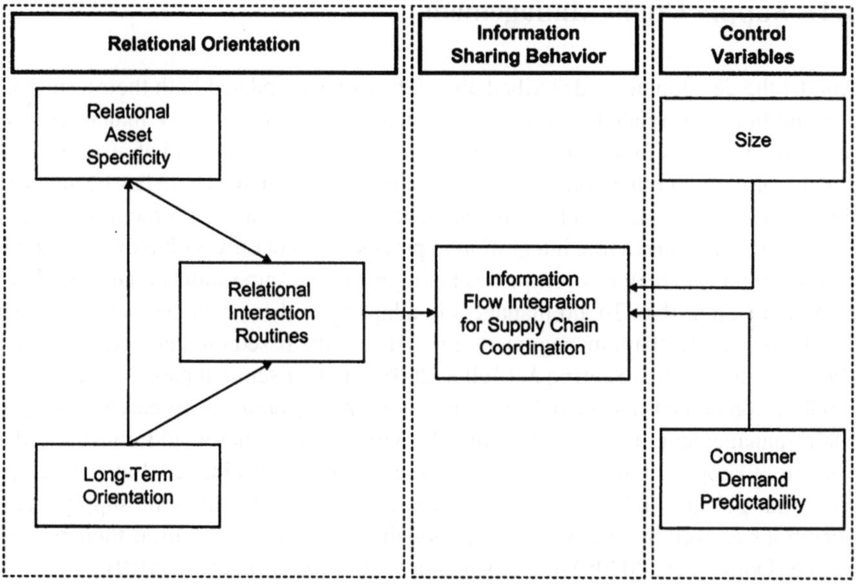

Fig. 1.2 Patnayakuni et al. (2006) research model. Source: Patnayakuni et al. (2006, p. 23)

manufacturing stages (Simchi-Levi et al., 1999). Thanks to advances in technology and information systems, firms are collecting data of broader scope and greater complexity in order to improve supply chain flows and the integration of information systems (Gunasekaran & Ngai, 2004; Turner et al., 2018). Of course, firms also need to cultivate their propensity and methods for interpreting and sharing voluminous data with supply chain partners. The most significant aspect for understanding the measurement of impacts is the emphasis placed on the relational and necessarily bidirectional nature of the data-sharing process, as well as the micro-level propensity for data sharing. Patnayakuni et al. (2006, p. 23) developed and validated a framework (Fig. 1.2) to capture these elaborated dynamics.

Subsequent work has advanced the frontier by foregrounding information visibility and spatial reach as prerequisites for sustainable supply chain management. Busse et al. (2017) show that firms must match information-processing capacity to sustainability-related uncertainties that arise precisely because supply bases are global and opaque. Managers have pursued a partial solution by validating the use and diffusion of the digital traceability tools (i.e., blockchain) to link provenance, environmental metrics, and social compliance across dispersed nodes (Saberi et al., 2019; Yu et al., 2025). Consequently, more recent studies have specified the data objects, stakeholder flows, and perspectives needed to make sustainability information meaningful across geographically distributed chains (Schäfer, 2023). By consolidating the complex activities and multiple actors in the supply chain, these scholarly developments form the basis for the emerging discipline of supply chain management.

1.2 Supply Chain Management

Amidst the developments described above, firms have updated both their terminology and functions related to supply chain management. Specifically, managers now need competencies in not only internal procurement and logistics but also external collaboration with other companies in the same supply chain. There has been a particular increase in tasks and interdependencies between a mere procurement and logistics function and a true integration of processes—not only within the individual company, but also between companies belonging to the same supply chain (Naslund & Williamson, 2010). To aid managers in adapting to these changes, scholars have developed supply chain management into a discipline based on their own literature (Janvier-James, 2012; Seuring & Müller, 2008). In their seminal paper, Croom et al. (2000, p. 68) noted that their "concern was with the nature of research in supply chain management, and more specifically with exactly what would constitute the domain of supply chain management as a management discipline." Later developments have clearly linked the discipline to the dynamics that underlie supply chain operations as well as to business activities that can optimize or limit their performance (Dong et al., 2020; Mukhamedjanova, 2020; Sharma et al., 2019).

According to (Harland, 1996, p. 64), "there are four main uses of the term 'supply chain management': 1. The internal supply chain integrates business functions involved in the flow of materials and information from the inbound to the outbound ends of the business. 2. Management of dyadic, or two-party, relationships with immediate suppliers. 3. The management of a chain of businesses, including a supplier, a supplier's suppliers, a customer, and a customer's customer. 4. The management of a network of interconnected businesses involved in the ultimate provision of product and service packages required by end customers." Clearly, companies remain focused on internal coordination to improve the efficiency of their (productive) processes, but the reference to dyadic and network dynamics underscores the plurality of external actors and the factors involved. This work particularly highlights the interactions within and between the firm, institutional, and supply chain levels, as well as the interaction between them (see Chap. 5 for the proposed theoretical framework). Gereffi et al. (2005) argue that governance patterns in global value chains hinge on three interlocking attributes of the buyer–supplier transaction: (i) information complexity (i.e., how difficult it is to communicate and coordinate the product and process requirements); (ii) codifiability (i.e., the degree to which that know-how can be standardized and efficiently transmitted without heavy, transaction-specific investment); and (iii) supplier capability (i.e., how well current or prospective suppliers can meet those requirements). When these three factors vary, distinct governance types emerge (Gereffi et al., 2005, pp. 84–85). Here, it is useful to revisit the five governance configurations originally outlined by Gereffi et al. (2005), which serve not only to describe structural patterns but also anticipate where and how sustainability data can (or cannot) be generated and shared. In their framework, governance outcomes follow the interplay of three factors: information complexity, the degree to which information can be codified, and supplier capabilities. Variations in these factors yield distinct coordination and power profiles across the five types (Gereffi et al., 2005, pp. 84–87). Figure 1.3 shows the five major types: market, modular, relational, captive, and hierarchical. Building on this foundation, early sustainability work in supply chains has shown that credible assessments must track material, information, and capital flows against all three sustainability dimensions (economic, environmental, and social) across the chain. These works highlight that most extant research was environmentally sensible and first-tier focused, leaving major data blind spots upstream (Seuring & Müller, 2008, pp. 1702–1703).

Viewed through a sustainability-data lens, each governance type presents different incentives and levers. Market structures (low complexity, high codification, capable suppliers) allow low-cost switching, but offer limited leverage to compel deeper ESG data unless demanded at transaction entry (Turner et al., 2018). Modular networks, where complex specs are codified and handed off digitally, are conducive to standardized sustainability data packets and lowering data collection costs across partners. Relational governance relies on dense, trust-based ties to exchange tacit know-how. These same high-touch interactions can surface nuanced social and environmental indicators if partners agree to invest in shared metrics. Captive chains, marked by powerful lead firms overseeing less capable suppliers, create strong top-down incentives (and tools) for mandated sustainability data capture, audits, and capacity building. Hierarchy internalizes both production and information, so firms can embed end-to-end sustainability measurement but may lack the

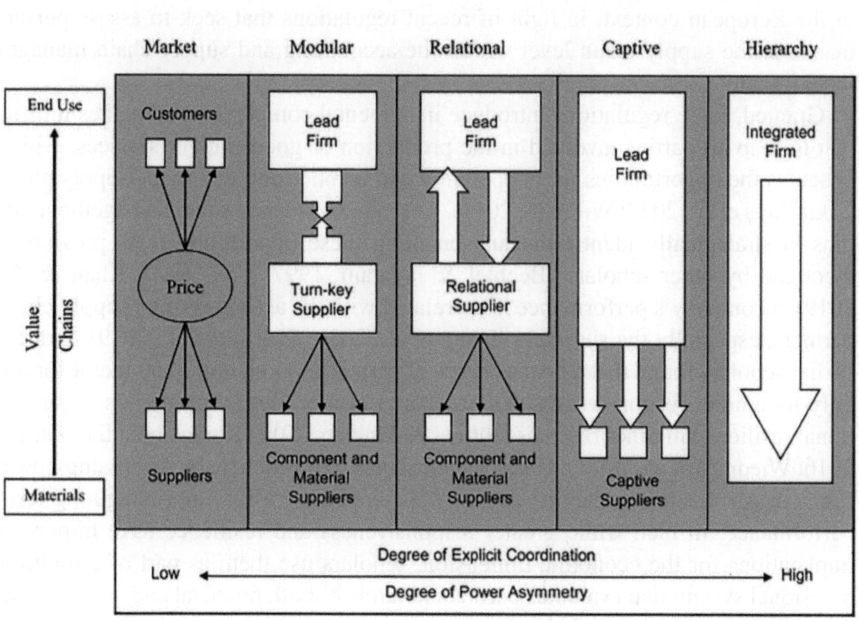

Fig. 1.3 Five global value chain governance types according to the Gereffi et al. (2005) framework. Source: Gereffi et al. (2005, p. 89)

benchmarking pressure that inter-firm disclosure regimes impose (Gereffi et al., 2005, pp. 85–86).

Subsequent empirical research confirms that multitier visibility—especially beyond tier-1—is essential because the greatest environmental and social risks often reside among lower-tier suppliers who remain data-poor and weakly monitored. Extending sustainability metrics upstream is thus not simply a technical challenge, but also a governance one (Villena & Gioia, 2018, pp. 66–68). This newer literature has given rise to technology-enabled transparency responses. Indeed, digital traceability architectures (i.e., blockchain layers and shared ledgers) are being deployed to bridge codification gaps and move verified sustainability attributes across dispersed value-chain nodes, reflecting an evolution from Beamon-era flow models toward data-rich, multi-party governance systems (Saberi et al., 2019, pp. 2119–2121; Yu et al., 2025).

More recently, transparency scholarship has unpacked what sustainability claims need to be credible: which data must travel (object), whose perspective defines adequacy (stakeholder), and across which tier the information flows. This agenda directly links governance configuration to the design of sustainability measurement and disclosure regimes (Schäfer, 2023, pp. 583–586). Research now underscores that multinational lead firms face expanding normative and regulatory expectations to diffuse sustainability practices (and the associated reporting process) across their global, multitier supply networks, making governance choices central to data scope, quality, and accountability (De Marchi & Gereffi, 2024). This book contributes to this debate by identifying the mechanisms that are guiding performance disclosures in the European context, in light of recent regulations that seek to assess performance at the supply chain level within the accounting and supply chain management literature.

Granted, these regulations introduce incremental complexity that makes it difficult to map all parties involved in the production of goods and/or services, which obscures the opportunities for cooperation and competition within the supply chain (Akın Ateş et al., 2022; Wilhelm, 2011). The value of supply chain management lies thus in strategically identifying and enabling these opportunities. As previously theorized by other scholars (Bechtel & Jayaram, 1997; Cox, 1999; Khan & Yu, 2019), a company's performance is correlated with the activities of its supply chain partners, especially the suppliers (Dong et al., 2020; Sharma et al., 2019). Indeed, earlier scholars found that a firm's choice of partners was an important factor for not only its market responsiveness (Gunasekaran et al., 2008), but also its organizational resilience in times of crisis (Jüttner & Maklan, 2011; Kamalahmadi & Parast, 2016; Wiedmer et al., 2021). Chap. 4 contributes to this debate by discussing how a firm's data-collection capability serves as a competitive asset, improving long-term performance. In fact, while greater responsiveness and resilience have important implications for the economic dimension, scholars use them as part of a multidimensional system that evaluates partners in terms of both financial and nonfinancial criteria (Sauer & Seuring, 2018).

1.3 Sustainable Supply Chain Management

This subsection aims to reconstruct the process of integrating the environmental (Tate et al., 2013) and social (Bals & Tate, 2018) dimensions into the criteria for selecting supply chain partners (i.e., suppliers) following the multidimensional approach proposed by Sauer and Seuring, 2018) or Wieland (2021). Moreover, this subsection contributes to the debate on data collection and performance evaluation in the supply chain context. Specifically, it reviews the prior research on sustainable supply chain management (left part of the proposed framework; see Sect. 5.1) and the theoretical basis for implementing successful supply chains in practice (e.g., Carter et al., 2015; Touboulic et al., 2020; see also the right part of the proposed framework).

As companies increasingly prioritize sustainability amidst mounting environmental concerns (Fahimnia et al., 2019), the supply chain management literature has focused on developing and promoting a strategic transition of production models to reduce the carbon footprint of manufacturing systems (Khan et al., 2021). Indeed, scholars of so-called green supply chain management are seeking to classify, define, implement, and advance a greater orientation toward sustainability, with the goal of shaping external pressures (e.g., regulatory standards) that can encourage the adoption of sustainable practices (Ahi & Searcy, 2013; Rajeev et al., 2017). A critical aspect of this transition is encouraging firms to integrate the triple-bottom-line (TBL) approach (Mardani et al., 2020; Tate & Bals, 2018). TBL is also associated with an extensive body of SSCM literature and forms the basis of various reference models and regulations. TBL represents the birth of an extensive body of literature concerning sustainable supply chain management and all the reference models (theoretical, practical, and regulatory) that have followed and evolved over time (Beske & Seuring, 2014; Pagell & Shevchenko, 2014; Trianni et al., 2017). According to Seuring and Müller (2008), there are several forces that can incentivize sustainable supply chains, such as legal demands/regulations, customer demands of stakeholders, competitive advantage, environmental and social pressure groups, and reputation loss. Accordingly, these incentives can interact with both supporting factors (i.e., company-overlapping communication; management systems; monitoring, evaluation, reporting, sanctions; training education for purchasing employees and suppliers; integration of sustainability orientation into corporate policy) and barriers (i.e., higher costs; coordination complexity; inadequate supply chain communication). Figure 1.4 graphically aligns those supporting and hindering factors in relation to the corporate effort to evaluate suppliers and (co-)develop 'sustainable' products.

Other scholars have gone further by identifying endogenous and exogenous factors that affect the diffusion of sustainable supply chain management (Galli, et al., 2024); Seuring et al., 2022; Wong et al., 2011). For instance, Gualandris and Kalchschmidt (2014) noted that market dynamics and stakeholder expectations are significant drivers of this transition, as are corporate leaders' commitment to sustainability and implementation of internal strategies. Scholars have identified a lot

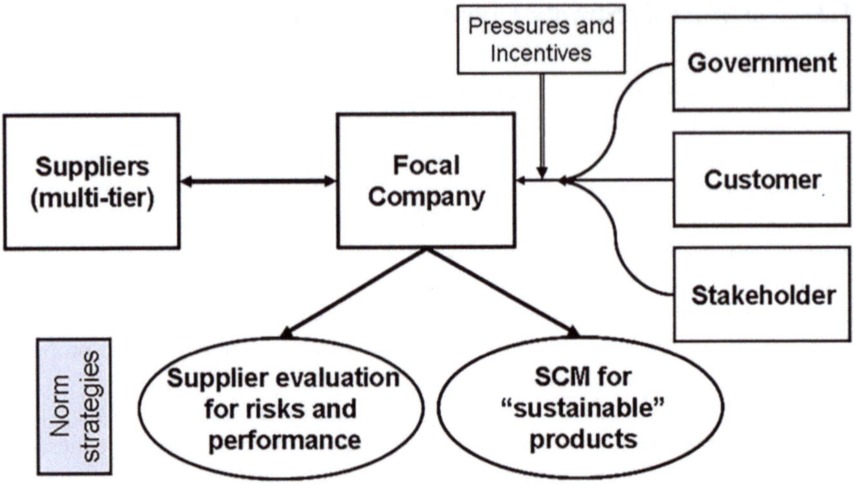

Fig. 1.4 Seuring and Müller's (2008) framework. Source: Seuring & Müller (2008, p.1703)

of these drivers or barrier factors; the next step is defining the guidelines and best practices for orienting a supply chain toward sustainability.

By broadening the focus to the value chain and its distribution among the various subjects belonging to the supply and production chain, this literature has significantly broadened not only the scope of inquiry but also the grounds for intervention (Khalid et al., 2015; Mardani et al., 2020; see Sect. 1.4 and 1.5). The key to interpreting and practically implementing the foundational elements lies, therefore, in the practical definition of guidelines, recommendations, and best practices—both in terms of streamlining supplies and in selecting suppliers and distribution channels best suited to embracing their distinctive orientation toward product and process sustainability (Bastas & Liyanage, 2018; Luthra et al., 2017). In this vein, Wu et al. (2014) advocated for the triangulation between supply chain strategy, corporate environmental strategy, and firm performance. Their findings (although limited to the Taiwanese context) reinforced the positive link between environmental sustainability actions and performance improvement, including economic performance. Further studies, such as Liu et al. (2023) in China, Creazza et al. (2023) in Italy, and Silva et al. (2021), have confirmed and extended the validity of these assumptions in Brazil.

Granted, many of these scholarly insights underscore that sustainability is a strategic orientation and does not erase the power dynamics in the production chain. That said, there would still be value in understanding how sustainable supply chain management alters the intensity of competitive pressure (Khalid & Seuring, 2019).

1.4 The Role of Sustainable Supply Chain Practices and Performance

Firms seeking to enhance their sustainability performance are increasingly focusing on sustainable supply chain management (SSCM). In this regard, Sajjad et al. (2020) explored the drivers of and barriers to SSCM implementation. One major driver they identified is regulatory pressure: Governments and international bodies are increasingly imposing regulations that mandate sustainable practices (Carter & Rogers, 2008). Through their compliance with these regulations, firms not only avoid legal repercussions but also benefit from improved reputations and operational efficiencies. For instance, businesses that adhere to environmental regulations can reduce their waste and energy consumption, which directly impacts their sustainability performance (Zhu et al., 2012).

Market demand is also a strong driver: Consumers are becoming more environmentally conscious and demand products that are sustainably produced (Linton et al., 2007). Businesses that respond to this demand can gain a competitive advantage through sustainability credentials and a stronger market position. This market-driven approach encourages companies to innovate and adopt sustainable practices throughout their supply chains, from material sourcing to product delivery (Awan et al., 2021; Seuring & Müller, 2008).

The last major driver is stakeholder pressure imposed by investors, nongovernmental organizations (NGOs), and local communities (Meixell & Luoma, 2015; Peng et al., 2022). These entities are increasingly essential to holding businesses accountable for their environmental and social impacts.

These drivers are, of course, offset by several barriers that hinder the implementation of SSCM. One significant barrier is the perceived cost: Many managers believe that sustainable practices require substantial financial investments that may not yield immediate returns (Carter & Easton, 2011). This short-term financial outlook often prevents companies from adopting practices that could provide long-term sustainability benefits.

A second barrier is the complexity of supply chains (Serdarasan, 2013). Global supply chains involve multiple stakeholders, each with their own interests and practices (Simpson et al., 2007). Coordinating and aligning these diverse entities toward common sustainability goals can be challenging. The lack of transparency (Vachon & Klassen, 2006) and traceability (Garcia-Torres et al., 2019) in supply chains further complicates the implementation of sustainable practices.

Another barrier is the limited knowledge and expertise of many managers and supply chain professionals who may not be fully aware of the benefits or methods of implementing sustainable practices (Walker et al., 2008). This knowledge gap can hinder the development and execution of effective and sustainable supply chain management strategies.

Implementing sustainable supply chain practices can significantly enhance a company's sustainability performance; thus, companies looking to overcome these barriers need to consider multiple dimensions of sustainable performance. The first way is through improved environmental performance: By reducing emissions, minimizing waste, and conserving resources, firms can achieve a lower environmental footprint (Kalaitzi et al., 2018; Sarkis et al., 2011) while also increasing their value (Perdichizzi et al., 2024). To this end, companies can optimize their logistics and adopt green technologies and source materials from sustainable channels (Eltayeb et al., 2011). Second, SSCM can help firms enhance their social performance by ensuring fair labor practices, supporting local communities, and engaging in ethical sourcing (Carter & Jennings, 2002). These efforts not only enhance the company's reputation but also foster greater customer loyalty and employee motivation (Klassen & Vereecke, 2012). Lastly, SSCM can positively influence economic performance. Improved efficiency, reduced waste, and enhanced brand reputation can lead to cost savings and increased revenue in the long term (Rao & Holt, 2005). While the implementation of sustainable practices may involve initial costs, companies that invest in sustainability often find themselves better positioned to mitigate the risks associated with socio-environmental regulations (Wieland et al., 2023) and resource scarcity (Kalaitzi et al., 2018), leading to more resilient and profitable operations (Paulraj, 2011).

1.5 The Distribution of Value Along the Supply Chain

The distribution of value in the production chain is an important topic, both for defining the entities involved and for determining the balance of power among them (Cox, 1999; Grabs & Ponte, 2019; Hines & Rich, 1997). In his 1985 book "Competitive Advantage," Michael Porter introduced the concept of the 'value chain' and defined it as the sum of nine general activities that firms engage in to create value for customers. He further extended this concept to a 'value system' that reflected the link between different firms' value chains. His work inspired analyses of how firms create value internally, although scholars later broadened the impact of external factors (contextual and institutional) on ESG risk and competitive advantage. In this regard, the framework proposed by Adams (2017) and reported in Fig. 1.5 includes these dynamics and links them to both the reporting process and related governance aspects (see also Sect. 4.3).

Today, due to firms' increasing reliance on outsourcing and collaboration, this interconnected network of value-creating processes is commonly referred to as the 'value chain.' There is an emphasis in the value chain on customer benefits, interdependent processes, and the resulting demand and financial flows (Ramsay, 2005).

According to Mentzer et al. (2001), the supply chain can be viewed as an interconnected network of organizations, individuals, activities, information, and resources that collaborate to move a product or service from its origin to the final customer. These activities convert natural resources, raw materials, and components into finished products that are delivered to the end user. Through both upstream and downstream linkages, each entity involved contributes to generating value in the form of goods and services provided to the ultimate consumer.

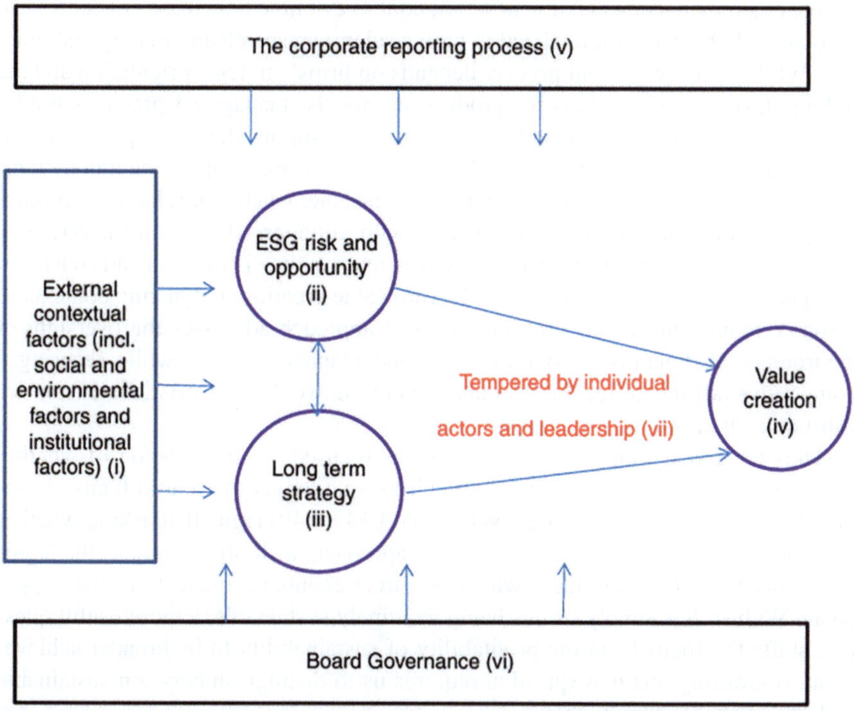

Fig. 1.5 Conceptualization of the corporate value creation process according to Adams (2017). Source: Adams (2017, p. 908)

In these early formulations, the connection between the implemented managerial practices and the direct benefits perceived by the consumer is evident. Al-Mudimigh et al. (2004) highlighted the need for a broader perspective that incorporates this connection within the value chain management literature. They found that supply chain management can optimize processes while noting its limitations in achieving holistic value creation. The authors thus introduced a "Value Chain Management" model focused on critical success factors, including integration, collaboration, information sharing, and technology utilization. This model centers effective value chain management on firms' agility in responding to market demands and customer needs. Later, scholars expanded their model in terms of its dimensions and practices (Beske & Seuring, 2014).

Sustainability reports have evolved to reflect this elaborate view of supply chains, making distinctions between the value generated within the company and value stemming from each stakeholder category (e.g., suppliers, customers, etc.) (Dumond, 2000). Therefore, the sustainability reporting process not only influences internal management control and optimizes supply chain efficiency (Feller et al., 2006) in terms of both financial and nonfinancial performance (Zeng et al., 2025), but also opens up new interconnections with other fields of research. Having provided a necessary introductory overview of how the literature defines "value chain" and "value chain management," it is now essential to examine how these concepts have subsequently been operationalized in both academic research and managerial practice. While the development process depends on firms' strategic orientation and the definition of transition plans for production models, managerial practices implemented at both the corporate and meso levels of the supply chain also play a critical role (Klassen & Vachon, 2003). The literature is clear that cooperation among multiple entities within the supply chain is indispensable, whether it relates to optimizing procedures and procurement criteria with suppliers (Caccialanza, 2022) or enhancing the sustainability attributes downstream in the distribution and consumption phases (Linton et al., 2007). By linking value creation in dynamic contexts to resource management, this multidimensional approach addresses the oversight of environmental dynamism, contingencies, and managerial roles while securing a competitive advantage for the firm and generating wealth and value for customers (Sirmon et al., 2007).

Thus far, much supply chain analysis has focused on the question of whether sustainability pays. To affirm that profitability is no longer the central focus of supply chain analysis, Pagell and Shevchenko (2014, p. 49) argue that asking whether sustainability pays is no longer the correct approach, as it often neglects the legitimate concerns of stakeholders who lack direct economic interests in the supply chain. With so few supply chains being genuinely sustainable, a more useful question shifts the focus from the profitability of sustainability to its broader achievement. Answering this new question requires us to distinguish between sustainable and unsustainable supply chains.

In order to realize these advantages, the field needs a better understanding of this phenomenon. Moreover, this concept evolved over time according to the context analyzed, and the literature traced this evolution. The linear dynamics and dyadic relationship between producer and transformer, or transformer and consumer of goods and services, diminish once visibility beyond the first tier of the supply chain is exceeded, and suppliers of suppliers begin to be mapped (Grabs et al., 2016). At this stage, the logic of one-to-one relationships breaks down; instead, a grid of involved actors must enhance the distribution dynamics within their network (Bacq & Aguilera, 2022). According to Pil and Holweg (2006, p. 72), organizations within a value-grid framework can enhance performance through several pathways: vertically, by exploring upstream or downstream opportunities in tiers adjacent to their existing value chain; horizontally, by capitalizing on similar tiers in multiple value chains; and diagonally, by adopting a more holistic view that spans various value chains and tiers to boost performance and minimize risk.

Globalized production has decoupled where value is created, captured, and borne in terms of social-environmental cost; without transparent mapping of value flows alongside physical flows, sustainability reporting risks masking distributional imbalances across tiers and geographies. Comparative work on commodity supply chains shows that current transparency efforts often illuminate downstream brands while leaving upstream producers (who frequently absorb the greatest land, labor, and environmental burdens) largely invisible. Thus, there is a need to track not just origins but also who captures rents along the chain (Gardner et al., 2019). Early sustainability governance scholarship has already warned that transparency would become central to value-chain accountability in an era of globalization and informational governance; such authors have argued that meaningful disclosure must include information relevant to power and distribution, not only environmental metrics (Mol, 2015). Subsequent reviews of global value chains explicitly call for research and reporting that connects governance to value creation and distribution across multi-country networks—the precise gap that distribution-aware transparency can fill (Kano et al., 2020; Marano et al., 2024). A power-sensitive global value chain perspective further links governance arrangements to the terms and location of value addition, distribution, and capture, thus highlighting why the disclosure of commercial terms, margins, and bargaining conditions matters for sustainability outcomes (Dallas et al., 2019). Evidence from labor-intensive sectors shows that voluntary CSR reporting without transparency in purchasing practices or wage pass-through rarely improves workers' livelihoods, reinforcing the case of distributional data in sustainability reports (LeBaron et al., 2022).

Increasingly, successful businesses are employing a multifaceted value-grid mindset to seize new opportunities and respond effectively to emerging challenges. Understanding the value distribution dynamics among various entities and different categories of stakeholders is essential for determining the ideal or actual distribution within the supply chain.

1.6 Ethical Distribution of Value in Supply Chains

As firms look to boost their sustainability performance, scholars have begun evaluating the importance of value being equitably and transparently distributed along the supply chain. To this end, academics have sought to define what a sustainable supply chain looks like relative to its baseline (Montecchi et al., 2021; Wieland & Handfield, 2010).

Alghababsheh et al. (2023) emphasized the critical importance of ethical value distribution in supply chains, arguing that the cornerstone of sustainable supply chain relationships is the fair treatment of all (categories of) stakeholders. Their review explored various dimensions of justice in supply chain interactions, including distributive, procedural, and interactional justice.

Distributive justice concerns the equitable allocation of resources and rewards among supply chain participants (e.g., fair remuneration and profit-sharing practices). An inequitable value distribution can lead to dissatisfaction, reduced motivation, and potential conflicts, ultimately undermining the efficiency and sustainability of the supply chain (Carter & Easton, 2011; Gualandris & Kalchschmidt, 2014).

Procedural justice refers to the fairness of the processes and methods used to make decisions within the supply chain. This involves transparent decision-making processes, consistent application of rules, and the involvement of all relevant stakeholders in decision-making. By ensuring procedural justice, supply chains can enhance trust and commitment among partners, thereby fostering a collaborative environment conducive to long-term success (Kim et al., 2016; Klassen & Vereecke, 2012).

Interactional justice focuses on the quality of interpersonal interactions within the supply chain, specifically in terms of respect, dignity, and provision of timely and adequate information. Positive interactions can strengthen relationships and build social capital, which affects supply chains' smooth functioning and resilience (Bai et al., 2019; Pagell & Wu, 2009; Sarkis et al., 2011).

Following this reasoning, Alghababsheh et al. (2023) argued that justice in supply chain relationships is both a moral imperative and a strategic necessity. They suggest that interdisciplinary approaches—combining insights from management, ethics, and supply chain logistics—are vital for advancing our understanding of justice in supply chains. By prioritizing distributive, procedural, and interactional justice, companies can create win-win situations that benefit all stakeholders involved. This approach not only addresses ethical concerns but also heightens the overall efficiency and sustainability of the supply chain. Likewise, scholars like Hadorn et al. (2006) and Pagell and Shevchenko (2014) noted that researchers can only advance sustainability by adopting multidisciplinary or transdisciplinary approaches that explicitly account for diverse stakeholders—ranging from governments and NGOs to local communities and the ecosystems in which supply chains operate. Inclusivity is essential to moving sustainability from a formal concept to a substantive strategy.

In conclusion, the dialogue surrounding the value chain underscored how an integrated approach—one that optimizes resources and ensures fair remuneration across all tiers—can enhance the overall performance. However, sustaining that progress requires the systematic and timely measurement of (sustainable) performances (Caccialanza, 2022), along with the transparent reporting of those metrics (Adams, 2004; Owen et al., 2001). Such practices enable organizations to estimate and communicate their true impact, thereby fostering more responsible decision-making. Moreover, as Ketchen Jr and Hult (2007) emphasized, the development of best practices that create win-win opportunities hinges on a fair distribution of production factors throughout the supply chain.

Establishing common standards for sustainability reporting would push this dialogue beyond its current boundaries, ensuring that accountability, competitive advantage, and stakeholder interests become integral elements of the entire value chain system.

References

Adams, C. A. (2004). The ethical, social, and environmental reporting-performance portrayal gap. *Accounting, Auditing & Accountability Journal, 17*(5), 731–757.

Adams, C. A. (2017). Conceptualising the contemporary corporate value creation process. *Accounting, Auditing & Accountability Journal, 30*(4), 906–931.

Ahi, P., & Searcy, C. (2013). A comparative literature analysis of definitions for green and sustainable supply chain management. *Journal of Cleaner Production, 52*, 329–341.

Akın Ateş, M., Suurmond, R., Luzzini, D., & Krause, D. (2022). Order from chaos: A meta-analysis of supply chain complexity and firm performance. *Journal of Supply Chain Management, 58*(1), 3–30.

Alghababsheh, M., Gallear, D., & Saikouk, T. (2023). Justice in supply chain relationships: A comprehensive review and future research directions. *European Management Review, 20*(3), 367–397.

Al-Mudimigh, A. S., Zairi, M., & Ahmed, A. M. M. (2004). Extending the concept of supply chain: The effective management of value chains. *International Journal of Production Economics, 87*(3), 309–320.

Awan, U., Arnold, M. G., & Gölgeci, I. (2021). Enhancing green product and process innovation: Towards an integrative framework of knowledge acquisition and environmental investment. *Business Strategy and the Environment, 30*(2), 1283–1295.

Bacq, S., & Aguilera, R. V. (2022). Stakeholder governance for responsible innovation: A theory of value creation, appropriation, and distribution. *Journal of Management Studies, 59*(1), 29–60.

Bai, C., Kusi-Sarpong, S., Badri Ahmadi, H., & Sarkis, J. (2019). Social sustainable supplier evaluation and selection: A group decision-support approach. *International Journal of Production Research, 57*(22), 7046–7067.

Bals, L., & Tate, W. L. (2018). Sustainable supply chain design in social businesses: Advancing the theory of supply chain. *Journal of Business Logistics, 39*(1), 57–79.

Bastas, A., & Liyanage, K. (2018). Sustainable supply chain quality management: A systematic review. *Journal of Cleaner Production, 181*, 726–744.

Beamon, B. (1998). Supply chain design and analysis: Models and methods. *International Journal of Production Economics, 55*(3), 281–294. https://doi.org/10.1016/S0925-5273(98)00079-6

Bechtel, C., & Jayaram, J. (1997). Supply chain management: A strategic perspective. *The International Journal of Logistics Management, 8*(1), 15–34.

Beske, P., & Seuring, S. (2014). Putting sustainability into supply chain management. *Supply Chain Management: An International Journal, 19*(3), 322–331.

Busse, C., Meinlschmidt, J., & Foerstl, K. (2017). Managing information processing needs in global supply chains: A prerequisite to sustainable supply chain management. *Journal of Supply Chain Management, 53*(1), 87–113. https://doi.org/10.1111/jscm.12129

Caccialanza, A. (2022). Supply chain management and procurement. In R. Brinkmann (Ed.), *The Palgrave handbook of global sustainability*. Palgrave Macmillan. https://doi.org/10.1007/978-3-030-38948-2_101-1

Carter, C. R., & Easton, P. L. (2011). Sustainable supply chain management: Evolution and future directions. *International Journal of Physical Distribution & Logistics Management, 41*(1), 46–62.

Carter, C. R., & Jennings, M. M. (2002). Logistics social responsibility: An integrative framework. *Journal of Business Logistics, 23*(1), 145–180.

Carter, C. R., Meschnig, G., & Kaufmann, L. (2015). Moving to the next level: Why our discipline needs more multilevel theorization. *Journal of Supply Chain Management, 51*, 94–102.

Carter, C. R., & Rogers, D. S. (2008). A framework of sustainable supply chain management: Moving toward new theory. *International Journal of Physical Distribution & Logistics Management, 38*(5), 360–387.

Choi, T. Y., & Hong, Y. (2002). Unveiling the structure of supply networks: Case studies in Honda, Acura, and DaimlerChrysler. *Journal of Operations Management, 20*(5), 469–493. https://doi.org/10.1016/S0272-6963(02)00025-6

Christopher, M. (1999). *Logistics and supply chain management: Strategies for reducing cost and improving service*. Financial Times: Pitman Publishing, London. ISBN 0273630490.

Cox, A. (1999). Power, value and supply chain management. *Supply Chain Management: An International Journal, 4*(4), 167–175.

Craighead, C. W., Blackhurst, J., Rungtusanatham, M. J., & Handfield, R. B. (2007). The severity of supply chain disruptions: Design characteristics and mitigation capabilities. *Decision Sciences, 38*(1), 131–156.

Creazza, A., Colicchia, C., & Evangelista, P. (2023). Leveraging shippers-logistics providers relationships for better sustainability in logistics: The perspective of SMEs. *The International Journal of Logistics Management, 35*(4), 1009.

Croom, S., Romano, P., & Giannakis, M. (2000). Supply chain management: An analytical framework for critical literature review. *European Journal of Purchasing & Supply Management, 6*(1), 67–83.

Dallas, M. P., Ponte, S., & Sturgeon, T. J. (2019). Power in global value chains. *Review of International Political Economy, 26*(4), 666–694.

De Marchi, V., & Gereffi, G. (2024). Multinational firms and sustainability in global supply chains: Scope and boundaries of responsibility. *Journal of International Business Studies, 55*, 413–428. https://doi.org/10.1057/s41267-024-00706-6

Dong, Y., Skowronski, K., Song, S., Venkataraman, S., & Zou, F. (2020). Supply base innovation and firm financial performance. *Journal of Operations Management, 66*(7–8), 768–796.

Dumond, E. J. (2000). Value management: An underlying framework. *International Journal of Operations & Production Management, 20*(9), 1062–1077.

Eltayeb, T. K., Zailani, S., & Ramayah, T. (2011). Green supply chain initiatives among certified companies in Malaysia and environmental sustainability: Investigating the outcomes. *Resources, Conservation and Recycling, 55*(5), 495–506.

Fahimnia, B., Sarkis, J., & Talluri, S. (2019). Editorial design and Management of Sustainable and Resilient Supply Chains. *IEEE Transactions on Engineering Management, 66*(1), 2–7. https://doi.org/10.1109/TEM.2018.2870924

Feller, A., Shunk, D., & Callarman, T. (2006). Value chains versus supply chains. *BP trends, 1*(3), 165–173.

Galli, D., Torelli, R., & Caccialanza, A. (2024). Sustainability performance and sustainability reporting in SMEs: a love affair or a fight? *Journal of Management & Organization, 30*(3), 574–599.

Garcia-Torres, S., Albareda, L., Rey-Garcia, M., & Seuring, S. (2019). Traceability for sustainability–literature review and conceptual framework. *Supply Chain Management: An International Journal, 24*(1), 85–106.

Gardner, T. A., Benzie, M., Börner, J., Dawkins, E., Fick, S., Garrett, R., et al. (2019). Transparency and sustainability in global commodity supply chains. *World Development, 121*, 163–177.

Gereffi, G., Humphrey, J., & Sturgeon, T. (2005). The governance of global value chains. *Review of International Political Economy, 12*(1), 78–104.

Grabs, J., Langen, N., Maschkowski, G., & Schäpke, N. (2016). Understanding role models for change: A multilevel analysis of success factors of grassroots initiatives for sustainable consumption. *Journal of Cleaner Production, 134*, 98–111.

Grabs, J., & Ponte, S. (2019). The evolution of power in the global coffee value chain and production network. *Journal of Economic Geography, 19*(4), 803–828.

Gualandris, J., & Kalchschmidt, M. (2014). Customer pressure and innovativeness: Their role in sustainable supply chain management. *Journal of Purchasing and Supply Management, 20*(2), 92–103.

Gunasekaran, A., Lai, K. H., & Cheng, T. E. (2008). Responsive supply chain: A competitive strategy in a networked economy. *Omega, 36*(4), 549–564.

Gunasekaran, A., & Ngai, E. W. (2004). Information systems in supply chain integration and management. *European Journal of Operational Research, 159*(2), 269–295.

Hadorn, G. H., Bradley, D., Pohl, C., Rist, S., & Wiesmann, U. (2006). Implications of transdisciplinarity for sustainability research. *Ecological Economics, 60*(1), 119–128.

Harland, C. M. (1996). Supply chain management: relationships, chains and networks. *British Journal of Management, 7*, S63–S80.

Hines, P., & Rich, N. (1997). The seven value stream mapping tools. *International Journal of Operations & Production Management, 17*(1), 46–64.

Janvier-James, A. M. (2012). A new introduction to supply chains and supply chain management: Definitions and theories perspective. *International Business Research, 5*(1), 194–207.

Jüttner, U., & Maklan, S. (2011). Supply chain resilience in the global financial crisis: An empirical study. *Supply Chain Management: An International Journal, 16*(4), 246–259.

Kalaitzi, D., Matopoulos, A., Bourlakis, M., & Tate, W. (2018). Supply chain strategies in an era of natural resource scarcity. *International Journal of Operations & Production Management, 38*(3), 784–809.

Kamalahmadi, M., & Parast, M. M. (2016). A review of the literature on the principles of enterprise and supply chain resilience: Major findings and directions for future research. *International Journal of Production Economics, 171*, 116–133. https://doi.org/10.1016/j.ijpe.2015.10.023

Kano, L., Tsang, E. W., & Yeung, H. W. C. (2020). Global value chains: A review of the multi-disciplinary literature. *Journal of International Business Studies, 51*(4), 577–622.

Ketchen, D. J., Jr., & Hult, G. T. M. (2007). Bridging organization theory and supply chain management: The case of best value supply chains. *Journal of Operations Management, 25*(2), 573–580.

Khalid, R. U., & Seuring, S. (2019). Analyzing base-of-the-pyramid research from a (sustainable) supply chain perspective. *Journal of Business Ethics, 155*, 663–686.

Khalid, R. U., Seuring, S., Beske, P., Land, A., Yawar, S. A., & Wagner, R. (2015). Putting sustainable supply chain management into base of the pyramid research. *Supply Chain Management: An International Journal, 20*(6), 681– 696.

Khan, S. A. R., & Yu, Z. (2019). *Strategic supply chain management* (pp. 1–290). Springer.

Khan, S. A. R., Yu, Z., Golpira, H., Sharif, A., & Mardani, A. (2021). A state-of-the-art review and meta-analysis on sustainable supply chain management: Future research directions. *Journal of Cleaner Production, 278*, 123357.

Kim, Y. H., Song, H. J., & Lee, J. (2016). The role of procedural justice in supply chain management: A social exchange perspective. *International Journal of Production Economics, 176*, 70–82.

Klassen, R. D., & Vachon, S. (2003). Collaboration and evaluation in the supply chain: The impact on plant-level environmental investment. *Production and Operations Management, 12*(3), 336–352.

Klassen, R. D., & Vereecke, A. (2012). Social issues in supply chains: Capabilities link responsibility, risk (opportunity), and performance. *International Journal of Production Economics, 140*(1), 103–115.

La Londe, B. J., & Masters, J. M. (1994). Emerging logistics strategies: Blueprints for the next century. *International Journal of Physical Distribution & Logistics Management, 24*(7), 35–47.

LeBaron, G., Edwards, R., Hunt, T., Sempéré, C., & Kyritsis, P. (2022). The ineffectiveness of CSR: Understanding garment company commitments to living wages in global supply chains. *New Political Economy, 27*(1), 99–115.

Linton, J. D., Klassen, R., & Jayaraman, V. (2007). Sustainable supply chains: An introduction. *Journal of Operations Management, 25*(6), 1075–1082.

Liu, Y., Fang, W., Feng, T., & Xi, M. (2023). Environmental strategy, green supply chain integration and sustainable performance: Examining the synergistic effects. *Management Decision, 61*(9), 2603–2628.

Luthra, S., Govindan, K., Kannan, D., Mangla, S. K., & Garg, C. P. (2017). An integrated framework for sustainable supplier selection and evaluation in supply chains. *Journal of Cleaner Production, 140*, 1686–1698.

Marano, V., Wilhelm, M., Kostova, T., Doh, J., & Beugelsdijk, S. (2024). Multinational firms and sustainability in global supply chains: Scope and boundaries of responsibility. *Journal of International Business Studies, 55*(4), 413–428.

Mardani, A., Kannan, D., Hooker, R. E., Ozkul, S., Alrasheedi, M., & Tirkolaee, E. B. (2020). Evaluation of green and sustainable supply chain management using structural equation modelling: A systematic review of the state of the art literature and recommendations for future research. *Journal of Cleaner Production, 249*, 119383.

Meixell, M. J., & Luoma, P. (2015). Stakeholder pressure in sustainable supply chain management: A systematic review. *International Journal of Physical Distribution & Logistics Management, 45*(1/2), 69–89.

Mentzer, J. T., DeWitt, W., Keebler, J. S., Min, S., Nix, N. W., Smith, C. D., & Zacharia, Z. G. (2001). Defining supply chain management. *Journal of Business Logistics, 22*(2), 1–25.

Mol, A. P. J. (2015). Transparency and value chain sustainability. *Journal of Cleaner Production, 107*, 154–161. https://doi.org/10.1016/j.jclepro.2013.11.012

Montecchi, M., Plangger, K., & West, D. C. (2021). Supply chain transparency: A bibliometric review and research agenda. *International Journal of Production Economics, 238*, 108152.

Mukhamedjanova, K. A. (2020). Concept of supply chain management. *Journal of Critical Reviews, 7*(2), 759–766.

Naslund, D., & Williamson, S. (2010). What is management in supply chain management?-a critical review of definitions, frameworks and terminology. *Journal of Management Policy and Practice, 11*(4), 11–28.

Oliver, R. K., & Webber, M. D. (2012). Supply-chain management: Logistics catches up with strategy. In *The roots of logistics* (pp. 183–194). Springer.

Owen, D. L., Swift, T., & Hunt, K. (2001). Questioning the role of stakeholder engagement in social and ethical accounting, auditing and reporting. *Accounting Forum, 25*(3), 264–282.

Pagell, M., & Shevchenko, A. (2014). Why research in sustainable supply chain management should have no future. *Journal of Supply Chain Management, 50*(1), 44–55.

Pagell, M., & Wu, Z. (2009). Building a more complete theory of sustainable supply chain management using case studies of 10 exemplars. *Journal of Supply Chain Management, 45*(2), 37–56.

Patnayakuni, R., Rai, A., & Seth, N. (2006). Relational antecedents of information flow integration for supply chain coordination. *Journal of Management Information Systems, 23*(1), 13–49.

Paulraj, A. (2011). Understanding the relationships between internal resources and capabilities, sustainable supply management and organizational sustainability. *Journal of Supply Chain Management, 47*(1), 19–37.

Peng, S., Jia, F., & Doherty, B. (2022). The role of NGOs in sustainable supply chain management: A social movement perspective. *Supply Chain Management: An International Journal, 27*(3), 383–408.

Perdichizzi, S., Buchetti, B., Cicchiello, A. F., & Dal Maso, L. (2024). Carbon emission and firms' value: Evidence from Europe. *Energy Economics, 131*, 107324.

Pettit, T. J., Croxton, K. L., & Fiksel, J. (2019). The evolution of resilience in supply chain management: A retrospective on ensuring supply chain resilience. *Journal of Business Logistics, 40*(1), 56–65.

Pienaar, W. (2009). *Introduction to business logistics*. Oxford University.

Pil, F. K., & Holweg, M. (2006). Evolving from value chain to value grid. *MIT Sloan Management Review, 47*(4), 72–80.

Rajeev, A., Pati, R. K., Padhi, S. S., & Govindan, K. (2017). Evolution of sustainability in supply chain management: A literature review. *Journal of Cleaner Production, 162*, 299–314.

Ramsay, J. (2005). The real meaning of value in trading relationships. *International Journal of Operations & Production Management, 25*(6), 549–565.

Rao, P., & Holt, D. (2005). Do green supply chains lead to competitiveness and economic performance? *International Journal of Operations & Production Management, 25*(9), 898–916.

Saberi, S., Kouhizadeh, M., Sarkis, J., & Shen, L. (2019). Blockchain technology and its relationships to sustainable supply chain management. *International Journal of Production Research, 57*(7), 2117–2135. https://doi.org/10.1080/00207543.2018.1533261

Sajjad, A., Eweje, G., & Tappin, D. (2020). Managerial perspectives on drivers for and barriers to sustainable supply chain management implementation: Evidence from New Zealand. *Business Strategy and the Environment, 29*(2), 592–604. https://doi.org/10.1002/bse.2389

Sarkis, J., Zhu, Q., & Lai, K. H. (2011). An organizational theoretic review of green supply chain management literature. *International Journal of Production Economics, 130*(1), 1–15.

Sauer, P. C., & Seuring, S. (2018). A three-dimensional framework for multi-tier sustainable supply chain management. *Supply Chain Management: An International Journal, 23*(6), 560–572.

Schäfer, N. (2023). Making transparency transparent: A systematic literature review to define and frame supply chain transparency in the context of sustainability. *Management Review Quarterly, 73*, 579–604. https://doi.org/10.1007/s11301-021-00252-7

Serdarasan, S. (2013). A review of supply chain complexity drivers. *Computers & Industrial Engineering, 66*(3), 533–540. https://doi.org/10.1016/j.cie.2012.12.008

Seuring, S., Aman, S., Hettiarachchi, B. D., de Lima, F. A., Schilling, L., & Sudusinghe, J. I. (2022). Reflecting on theory development in sustainable supply chain management. *Cleaner Logistics and Supply Chain, 3*, 100016.

Seuring, S., & Müller, M. (2008). From a literature review to a conceptual framework for sustainable supply chain management. *Journal of Cleaner Production, 16*(15), 1699–1710.

Sharma, A., Kumar, V., Yan, J., Borah, S. B., & Adhikary, A. (2019). Understanding the structural characteristics of a firm's whole buyer–supplier network and its impact on international business performance. *Journal of International Business Studies, 50*(3), 365–392.

Silva, M. E., Silvestre, B. S., Ponte, R. C. D. V., & Cabral, J. E. O. (2021). Managing micro and small enterprise supply chains: A multi-level approach to sustainability, resilience and regional development. *Journal of Cleaner Production, 311*, 127567.

Simchi-Levi, D., Kaminsky, P., & Simchi-Levi, E. (1999). *Designing and managing the supply chain: Concepts, strategies, and cases*. McGraw-hill.

Simpson, D., Power, D., & Samson, D. (2007). Greening the automotive supply chain: A relationship perspective. *International Journal of Operations & Production Management, 27*(1), 28–48.

Sirmon, D. G., Hitt, M. A., & Ireland, R. D. (2007). Managing firm resources in dynamic environments to create value: Looking inside the black box. *Academy of Management Review, 32*(1), 273–292.

Tate, W. L., & Bals, L. (2018). Achieving shared triple bottom line (TBL) value creation: Toward a social resource-based view (SRBV) of the firm. *Journal of Business Ethics, 152*, 803–826.

Tate, W. L., Ellram, L. M., & Gölgeci, I. (2013). Diffusion of environmental business practices: A network approach. *Journal of Purchasing and Supply Management, 19*(4), 264–275.

Touboulic, A., McCarthy, L., & Matthews, L. (2020). Re-imagining supply chain challenges through critical engaged research. *Journal of Supply Chain Management, 56*, 36–51.

Trianni, A., Cagno, E., & Neri, A. (2017). Modelling barriers to the adoption of industrial sustainability measures. *Journal of Cleaner Production, 168*, 1482–1504. https://doi.org/10.1016/j.jclepro.2017.07.244

Turner, N., Aitken, J., & Bozarth, C. (2018). A framework for understanding managerial responses to supply chain complexity. *International Journal of Operations & Production Management, 38*, 1433–1466.

Vachon, S., & Klassen, R. D. (2006). Extending green practices across the supply chain: The impact of upstream and downstream integration. *International Journal of Operations & Production Management, 26*(7), 795–821.

Villena, V. H., & Gioia, D. A. (2018). On the riskiness of lower-tier suppliers: Managing sustainability in supply networks. *Journal of Operations Management, 64*, 65–87. https://doi.org/10.1016/j.jom.2018.09.004

Walker, H., Di Sisto, L., & McBain, D. (2008). Drivers and barriers to environmental supply chain management practices: Lessons from the public and private sectors. *Journal of Purchasing and Supply Management, 14*(1), 69–85.

Wiedmer, R., Rogers, Z. S., Polyviou, M., Mena, C., & Chae, S. (2021). The dark and bright sides of complexity: A dual perspective on supply network resilience. *Journal of Business Logistics, 10*, 1–24. https://doi.org/10.1111/jbl.12264

Wieland, A. (2021). Dancing the supply chain: Toward transformative supply chain management. *Journal of Supply Chain Management, 57*(1), 58–73.

Wieland, A., & Handfield, R. B. (2010). The socially responsible supply chain: An imperative for global corporations. *Supply Chain Management Review, 14*(6), 22.

Wieland, A., Stevenson, M., Melnyk, S. A., Davoudi, S., & Schultz, L. (2023). Thinking differently about supply chain resilience: What we can learn from social-ecological systems thinking. *International Journal of Operations & Production Management, 43*(1), 1–21.

Wilhelm, M. M. (2011). Managing coopetition through horizontal supply chain relations: Linking dyadic and network levels of analysis. *Journal of Operations Management, 29*(7–8), 663–676.

Wong, C. W., Lai, K. H., & Cheng, T. C. E. (2011). Value of information integration to supply chain management: Roles of internal and external contingencies. *Journal of Management Information Systems, 28*(3), 161–200.

Wu, T., Wu, Y. C. J., Chen, Y. J., & Goh, M. (2014). Aligning supply chain strategy with corporate environmental strategy: A contingency approach. *International Journal of Production Economics, 147*, 220–229.

Yu, J., Yuan, G., Liu, S., & Zhang, J. (2025). Green credit, supply chain concentration and corporate digital transformation: Evidence from China. *The British Accounting Review*, 101690.

Zeng, Y., Katsumata, S., Li, X., & Zhang, Q. (2025). Supplier–buyer (in) congruence in environmental management accounting for sustainable development in the context of Japan. *Sustainable Development, 33*(1), 84–99.

Zhu, Q., Sarkis, J., & Lai, K. H. (2012). Green supply chain management innovation diffusion and its relationship to organizational improvement: An ecological modernization perspective. *Journal of Engineering and Technology Management, 29*(1), 168–185.

Chapter 2
The Antecedents of Supply Chain Accountability: The Role of Accounting and Accountability Principles at the Firm-Level

2.1 Stakeholders' Accountability and Expectations

As Sinclair (1995) observed, accountability is a complex and ambiguous concept. Various disciplines maintain their own views on what being accountable entails (Gray, 2001, 2013; Shearer, 2002). According to Gray et al. (1996), accountability can be defined as "the duty to provide an account of actions for which one is responsible" (p. 38), which is not strictly limited to firms' financial performance (Cooper & Owen, 2007). In this vein, it is important to define the actors who receive said account: These could be shareholders, stakeholders, or even society in a broad sense (Benston, 1982).

All external stakeholders need to be informed about the company's activities. As the variety of stakeholders expands, organizations need to refine their performance measurement and accounting systems to meet sustainability goals (Allais et al., 2017; Costa & Pesci, 2016; Dal Maso et al., 2017, 2018). In their seminal work, Hahn and Kühnen (2013) effectively demonstrated the connection between sustainability orientation (and its degree of integration) and internal sustainability accounting. This connection depends on the managerial practices implemented (Adams & Frost, 2008; Grabs & Garrett, 2023; Pérez-López et al., 2015; Winn & Angell, 2000) and the reporting tools adopted (Baumgartner & Ebner, 2010). Figure 2.1 reports Hahn and Kühnen's (2013) research framework, which illustrates the intercorrelations between three focal elements: sustainability orientation, the performance measurement established, and the respective reporting activities. The lower portion of Fig. 2.1, in particular, has implications for accountability at both the firm and meso levels. Specifically, historical evolution has led to an expansion of the considered dimensions and assessments of multiple focal areas (whether simultaneously or through integrated reporting tools). The relationship is therefore bidirectional: As the extent of sustainability considerations expands—moving from isolated

A. Caccialanza, *Sustainability and Accountability of Supply Chains*, SIDREA Series in Accounting and Business Administration,
https://doi.org/10.1007/978-3-032-16879-5_2

Fig. 2.1 Overview and relations of basic concepts and terminology relating to sustainability reporting as proposed by Hahn and Kühnen (2013). Source: Hahn & Kühnen (2013, p. 7)

dimensions to a holistic approach—organizations also adopt reporting practices that are increasingly comprehensive and informative.

Subsequent scholars expanded their model to clarify the impact of specific governance and institutional factors, as well as the maturity of firms' sustainability orientation (Appannan et al., 2023), relevant institutional factors (see Sect. 2.3), and specific governance factors (see also Sect. 4.3).

The process of sustainability integration typically starts by realigning the company's core activities and then expanding them to peripheral elements (non-core activities or evaluation of externalities), which typically entails the formalization of processes or certifications (Baumgartner & Ebner, 2010). Numerous scholars have attempted to systematize or model this process (Shukla & Adil, 2021). Some authors initially defined a linear progression (ideal) of adoption (Roberts & Gehrke, 1996) and then defined the steps and phases that companies can incrementally undertake (Potrich et al., 2019). These definitions are typically accompanied by descriptions of commonly associated managerial practices and, in some cases, empirical case studies for validation (Azzone & Noci, 1998). A consistent feature of these models is the progressive involvement of different stakeholder categories in the evaluation process (Potrich et al., 2019; Shukla & Adil, 2021). After all, active stakeholder engagement is integral to the evaluation and reporting of sustainable outcomes (Aureli et al., 2020; Baah et al., 2021). Including stakeholders in the accounting process itself also enhances the legitimacy of economic activity (Bansal & Clelland, 2004; Bebbington et al., 2017). Several studies have linked this crucial factor to corporate performance (Al-Tuwaijri et al., 2004; Cho & Patten, 2007; see also Sect. 4.2).

Consequently, the literature has expanded the number of stakeholder categories considered in the accountability process of defining and evaluating impacts in reporting processes (Parmigiani et al., 2011) or in the verifiability of performance disclosures (Rhianon Edgley et al., 2010). However, scholars have also debated the legitimacy of adopting standards that focus just on the needs of internal managers or (external) users versus the needs of all stakeholders and society at large (Boiral, 2016; Cormier et al., 2004). Consequently, decision-makers and leaders often base their choice of standards on cost-benefit analyses that neglect broader stakeholder interests. To enable effective accountability, firms need to engage multiple stakeholders in dialogue through dialogical accountability perspectives and the proper application of standards principles (Adams & McNicholas, 2007; Rasche & Esser, 2006). Granted, determining what is accountable depends not only on the audience but also on the reporting standards used in different sectors (Göbbels & Jonker, 2003, p. 54). Understanding these standards requires expanding from a micro perspective (the individual organization; Moggi et al., 2015) to the meso level (i.e., the supply chain). Indeed, the prominent TBL perspective (Elkington, 1998) has long emphasized the need to integrate the disciplines of supply chain management and accounting (Boone et al., 2012). For example, Foran et al. (2005) sought to unify a wide range of sustainability reporting initiatives within an integrated framework. They achieved this by designing a conceptual diagram that positions both consuming and producing entities alongside environmental assessments, monitoring methods, and philosophies employed in previous studies (Foran et al., 2005, p. 147).

Beyond the audience and sector, another key aspect of accountability stems from the sociocultural expectations placed on organizations (Adams et al., 1998; Woodward et al., 1996), especially during crises (Pesci et al., 2025). In this regard, scholars typically delineate accountability into contractual and non-contractual societal elements using stakeholder theory to illustrate these legitimacy relationships. For instance, Gualandris et al. (2015) uncovered evidence that both perceived and expected stakeholder accountability are relevant in sustainable evaluation and verification within supply chains. These dimensions of accountability are essential for aligning organizational practices with stakeholder expectations and enhancing overall performance and legitimacy in supply chain operations (Beske-Janssen et al., 2015; Lamming et al., 2005). Consequently, their study usefully advances the debate on stakeholder accountability by aligning firm-level sustainability assessments with meso-level dynamics. Specifically, their framework (reported in Fig. 2.2) combines the 'supply chain integration' with 'sustainable evaluation and verification' and 'perceived'/'expected' stakeholders' accountability. Moreover, the direction of the relation suggests that supply chain integration, firm capabilities, and stakeholder salience influence sustainable evaluation and verification. In Chap. 5, we further discuss the bidirectional nature of this influence within our theoretical framework (for further details, see Sect. 5.2, Accountability of the Value Chain).

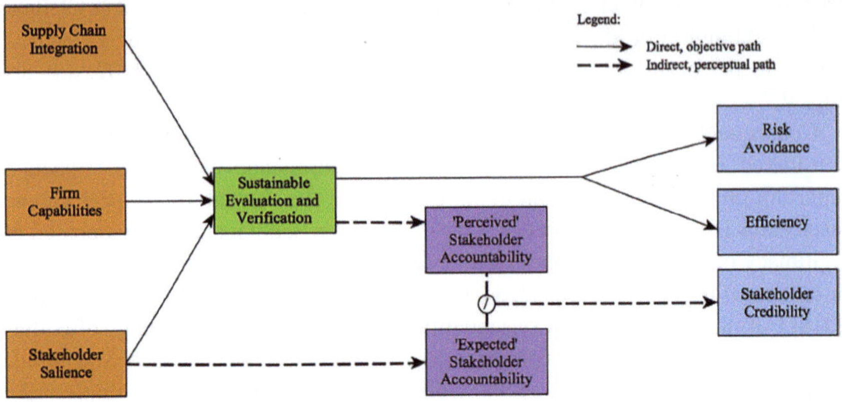

Fig. 2.2 Framework for the alignment and leveraging of accountability to stakeholders. Source: Adapted from Gualandris et al. (2015, p. 8)

At the corporate level, international reporting standards have raised awareness and promoted sustainability through the application of common sustainability metrics and shared reporting processes. The following subsection discusses these two standards in detail.

2.2 Disclosure Standards: Integrated Reporting and GRI

Performance measurement and sustainability disclosure can be costly activities for companies, particularly for those small- to medium-sized enterprises (SMEs) that lack robust structures (Caccialanza & Torelli, 2024; Galli et al., 2023). Thus, spreading greater awareness about performance improvement depends on identifying the best practices and tools (Azzone et al., 1996; Epstein, 2007), as well as diffusing standards such as integrated reporting (IR) or the Global Reporting Initiative (GRI). This subsection retraces the evolution of both standards, with a focus on their consequences for supply chain management. An overview of studies that focus on earlier sustainability reporting standards is relevant for at least three reasons. First, it examines the outcomes of applying other historically widespread standards, thereby enabling future research to compare performance effects under two distinct frameworks. Second, it allows us to assess the impacts on accountability achieved at the supply chain (meso-)level using standards that do not treat supply chain performance as the primary or prevailing unit of analysis but rather evaluate corporate externalities on the firm's value chain or in the broader operating context. Third, it clearly underscores how the analytical lens of supply chain accountability has remained limited in the literature under previous standards. Taken together, these three contributions synthesize prior views on supply chain accountability. In Chap. 5, we analyze the potential of the new standard from an incremental and/or comparative perspective.

2.2.1 *Integrated Reporting—IR*

Integrated reporting (IR) has evolved significantly over the last decade, moving from traditional financial reporting to a more inclusive approach that considers environmental, social, and governance factors (e.g., IFRS Foundation, 2021; IFRS Foundation Consultation Paper, 2020; IFRS-CF, 2018; IFRS-MC-ED, 2021). This evolution reflects a broader understanding of value creation and the increasing relevance of sustainable practices in the corporate sector (Dragu, 2019). The related literature features diverse perspectives on the impact of this evolution, but there is a common view that the field needs to expand its frameworks and methodologies in order to fully integrate sustainability into corporate reporting and operations, particularly within supply chains.

Brown and Dillard (2014) argued that IR should move beyond merely combining financial and non-financial data to instead include social and environmental dimensions (i.e., those related to sustainability). In support of this view, Stubbs and Higgins (2014) discussed how IR can serve as a mechanism for internal change, pushing corporate governance structures to align more closely with sustainable practices. More recently, Busco and Sofra (2021) explored the evolution of sustainability reporting by linking IR to sustainable development challenges. They suggested that current IR frameworks need to be further refined in order to respond to global sustainability challenges. In line with this reasoning, Lai and Stacchezzini (2021) produced a theoretical framework for understanding the organizational and professional challenges associated with the evolution of sustainability reporting. They also proposed an agenda for future research focused on the integration of sustainability into organizations' core strategic functions, which could facilitate more effective implementation and reporting of sustainable practices.

Several authors have investigated the factors that affect IR adoption and diffusion. Robertson and Samy (2015), for instance, addressed this topic using data on the UK FTSE 100 companies. They found that while regulatory support and stakeholder pressure are critical, companies' internal commitment to and understanding of IR's benefits also play a significant role. Oktorina et al. (2022) conducted a similar analysis on the determinants of IR adoption, but from a cross-country perspective. They highlighted that cultural, regulatory, and market factors influence how and why companies choose to adopt IR practices. Hossain et al. (2023) reviewed the literature on IR diffusion and identified several key factors. Their work suggests that despite the growing implementation of IR, there are significant gaps in understanding its impact on actual sustainability outcomes.

Although most of the contributions on IR adoption have focused on its internal impact on organizations, a few have weighed the broader impacts on external entities. Adams and Frost (2008) discussed the integration of sustainability reporting into management practices, including supply chain operations. They suggested that sustainability should be embedded into every level of supply chain management (i.e., from procurement to distribution) to ensure that sustainability is not just reported but actively managed and improved. Adams (2015, 2017) pushed further for the integration of Sustainable Development Goals (SDGs) into reporting practices. She argued that IR efficacy depends on "integrated thinking," which involves considering the interconnections between various sustainability issues and business operations, particularly in supply chains. Lastly, Perego et al. (2016) critiqued the current state of IR as having "a lot of icing but little cake" (p. 55). Their work suggested that despite academics' extensive investments in frameworks and discussions, firms have not committed to substantive changes (p. 55). The authors proposed that future developments should focus on actionable changes, particularly in how companies manage and report on supply chain sustainability. This book engages with this proposal by focusing on the application of sustainability standards to evaluate supply chain sustainability in accordance with new European standards, which are expected to be adopted far more widely than the IR framework in this geographical context.

2.2.2 Global Reporting Initiative—GRI

The Global Reporting Initiative (GRI) has significantly shaped the landscape of sustainability reporting, evolving from a guideline that emphasized transparency to a framework that is deeply integrated with corporate strategy. This subsection retraces the evolution of GRI standards, reflects on the insights provided by recent contributions, and identifies how these developments enhance sustainability practices, particularly in supply chains.

GRI was established in the late 1990s with the primary goal of bolstering global comparability and quality of sustainability reporting (Stranieri et al., 2019). GRI standards have evolved over the years, driven by the need for organizations to provide more meaningful and verifiable data that stakeholders demand. Antonini et al. (2020) highlighted that the GRI's shift toward G4 guidelines—and subsequently to GRI standards—reflects an increased focus on materiality and stakeholder inclusiveness, requiring firms to expand their reporting from the impact on business to the impact on society and the environment (Chen et al., 2015; Lee & Kim, 2009).

GRI standards have mainly been used for micro-level analyses (i.e., firms), but scholars have argued for their meso-level relevance (i.e., supply chains). For instance, Machado et al. (2023) examined how GRI standards can be practically used to reveal green supply chain management practices. They found that companies that adopt GRI reporting standards tend to have more organized and transparent supply chain practices. This observation aligns with earlier studies: For instance, Burritt and Schaltegger (2014) and Ahi and Searcy (2015) noted that firms increasingly leverage GRI reports to not only boost the transparency of their sustainability efforts but also encourage suppliers to adopt similar practices. Similarly, Fritz et al. (2017) emphasized that companies adopting GRI standards can better assess their compliance with environmental regulations and manage their operational risks more effectively. The success of these initiatives is linked not only to maximizing positive impacts but also to minimizing risks and conflicts. As confirmation of this, prior studies have already pointed out the role of auditors in managing this kind of risk and diminishing, for example, litigation (Dal Maso et al., 2020).

Despite these advancements, practitioners have struggled to fully integrate GRI standards into all aspects of supply chain management, thus limiting the applicability of these standards for supply chain analysis (Goharshenasan et al., 2022). Chen et al. (2015, p. 454) noted:

> … there are still some limitations for the GRI guideline itself. For example, GRI reports are not sufficient in providing information for the CSR performance assessment at suppliers′ level. As Lee and Kim (2009) discussed in their research, for company′s daily supply chain management practices, lack of specific information of the CSR performance assessment at suppliers′ level is a big disadvantage. The development of evaluation measures for suppliers can be a crucial determinant in improving a company′s performance. Therefore, it is very important for the GRI report developers considering how to develop concise and practical measurement indicators for the social performance from a supply chain perspective.

In the same vein, Boiral and Henri (2017) highlighted the need for greater accountability regarding the application of GRI standards and protocols. In their study of mining companies, the authors discovered a high level of data heterogeneity that rendered the results ambiguous. Their evidence underscored that certain sectors (those exposed to more scrutiny and scandals) are subject to more reporting opacity. In line with this reasoning, Fontoura and Coelho (2022) argued that GRI standards need to be more adaptable to diverse industries and supply chain structures to maximize their effectiveness. Thus, they suggested that future iterations of the framework should consider sector-specific issues more deeply to enhance their applicability and impact. Conversely, Chiarini (2017) noted that a significant number of companies that were actually reporting supply chain impacts or disclosing strict controls on their suppliers' sustainability were assemblers or firms that were not correlated with supply chains with severe environmental or social impacts. Relatedly, scholars such as Bunclark and Barcellos-Paula (2021) and Massari and Giannoccaro (2023) have highlighted that GRI needs to more directly address the sustainability impacts of emerging technologies and global trends such as digitalization and the circular economy. This integration could ensure that GRI standards continue to offer meaningful guidance in the face of new sustainability challenges (Moggi, 2023). For instance, digital technologies could be used to track and verify supply chain activities against GRI standards in real time and avoid the 'self-laudatory tone' that characterizes most of the sustainability reports (Holder-Webb et al., 2009).

On top of those considerations, to enhance the adoption of sustainability issues within supply chains, several new proposals have been considered. Fontoura and Coelho (2022) suggested adapting GRI standards to better fit different sectors and supply chain configurations. Similarly, utilize digital technologies to track and verify supply chain activities against GRI standards in real-time, improving the accuracy and timeliness of sustainability reporting (Long et al., 2023).

In summary, the evolution of the GRI framework has significantly contributed to the advancement of sustainability reporting and management, particularly within supply chains. By continuing to adapt to the changing business environment and emerging global challenges, GRI can help advance sustainable business practices. In a complementary fashion, this book fosters the debate about the applicability of reporting standards for assessing supply chain performances and impacts, while adding a third standard to the debate: the European Sustainability Reporting Standards (ESRS; for more details on this focus, see Sect. 3.3.3).

2.3 Scope of Disclosure

The scope of socio-environmental disclosure refers to the breadth and depth of information that companies provide regarding their environmental and social impacts. This includes data on greenhouse gas emissions, resource usage, social contributions, labor practices, and other activities that affect stakeholders. The scope of disclosure is critical because it determines how much information is available to stakeholders, thus influencing their ability to hold companies accountable for their actions (Okada et al., 2018).

The scope of socio-environmental disclosure has evolved significantly in recent decades: Initially, disclosures were minimal and often limited to regulatory requirements. However, driven by various stakeholders' growing awareness and pressure, companies have expanded their disclosures. This shift is evident in the increasing number of companies reporting Scope 3 emissions, which include indirect emissions from the entire value chain (Asif et al., 2022).

This evolution can be traced back to early environmental reports in the 1990s, which were primarily qualitative and lacked standardized metrics (Fekrat et al., 1996). With the advent of sustainability reporting frameworks like the Global Reporting Initiative (GRI) and the Carbon Disclosure Project (CDP), there has been a push toward more comprehensive and quantitative reporting. For example, transitioning economies (Gagalyuk et al., 2021) have broadened their CSR disclosure practices to include a wider range of social and environmental metrics.

Several factors have driven the expansion of socio-environmental disclosure, the strongest of which is regulatory pressure (see Sect. 2.3). Governments and international bodies have increasingly mandated more comprehensive disclosures, pushing companies to expand the scope of their reporting (Baldini et al., 2018; Karatzoudi & Aven, 2024). As a result, companies are motivated to improve their disclosures to attract and retain investments (Panjwani et al., 2023). The other critical drivers are public awareness and consumer demand for transparency. Consumers are becoming more informed and concerned about the environmental and social impacts of their purchases. As a result, companies are under pressure to provide detailed and accurate information (Hettler & Graf-Vlachy, 2024).

Despite these drivers, there are significant barriers to expanding socio-environmental disclosure. One major barrier is the lack of standardized reporting frameworks, which hinders companies' ability to know what and how to report. This issue is compounded by the varying requirements across different regions and industries (Qian & Schaltegger, 2017). Another barrier is the cost and resource intensity associated with comprehensive reporting. Gathering, verifying, and reporting detailed socio-environmental data can be expensive and time-consuming, particularly for smaller companies with limited resources (Yasar & Yalçın, 2023). Corporate culture and resistance to change also pose barriers: Some companies may be reluctant to disclose negative information or fear that increased transparency could lead to scrutiny and criticism (Cao et al., 2021).

It is worth highlighting the varying approaches that different sectors take to socio-environmental disclosure, reflecting their unique challenges and stakeholder expectations. For example, in the industrial sector, companies often focus on reporting emissions, energy usage, and waste management practices. In line with this reasoning, Suwaidan et al. (2004) found that Jordanian industrial companies tended to disclose information related to their environmental management systems and pollution control measures. The food and agribusiness sector, on the other hand, emphasizes disclosures related to supply chain sustainability and resource usage (Caccialanza & Torelli, 2024; Galli et al., 2023). Thus, companies in this sector are increasingly reporting on Scope 3 emissions to address the environmental impacts across their entire supply chain (Asif et al., 2022). In the technology and manufacturing sectors, there is a growing focus on disclosing information on product lifecycle impacts and resource efficiency. Kundeliene and Stepanauskaite (2018) observed that Lithuanian companies in these sectors are enhancing their online disclosures to provide more comprehensive information on their environmental and social impacts. The financial sector, which has traditionally limited socio-environmental disclosures, is now increasingly integrating ESG criteria into its reporting. Aldhizer III et al. (2009) discussed how financial firms are adopting more rigorous disclosure practices in response to investor demands for greater transparency.

These sector-specific approaches indicate that while there is a general trend toward more comprehensive socio-environmental disclosure, the specifics of what is disclosed and how it is reported can vary significantly based on the industry's characteristics and stakeholder expectations (Dietz et al., 2021).

2.4 Double Materiality

The evolving concept of materiality in sustainability reporting represents a significant transformation in how businesses communicate the flow of impact between their activities and operational environments (Mio et al., 2024). This analysis draws from the breadth of academic contributions to define materiality, chart its historical evolution and contextual variations, and explore its applications.

'Materiality' in the context of sustainability reporting refers to the significance of information concerning the economic, environmental, and social impacts of the organization (Dodsworth et al., 2023). It determines which issues are worthy of reporting based on their potential to influence stakeholders' decisions. Cooper and

Michelon (2022) defined materiality as "the threshold at which sustainability issues become sufficiently important that they should be reported" (p. 52). More specifically, 'financial materiality' relates to the impact of economic information that matters to users of financial statements. It is traditionally linked to what could affect a company's financial conditions or operational results (Boissinot et al., 2022). This aspect of materiality is often centered around the investor's perspective, focusing on the financial implications of sustainability issues (Edgley et al., 2015, p. 104). More recently, scholars have embraced the concept of double materiality to capture the financial impact of sustainability issues on not just the company but also society and the environment (e.g., Dragomir et al., 2025; Setia et al., 2024). This dual perspective is critical for understanding the reciprocal influences between a corporation and its external environment. For instance, Stroehle et al. (2022) described double materiality as "a framework that requires companies to consider both how sustainability issues affect their financial performance and how their operations impact sustainability," which underscores the bidirectional nature of the impact (p. 95).

In short, stakeholder materiality assessment is a process in sustainability reporting that determines which issues are of significant interest to both the organization and its stakeholders (Jørgensen et al., 2022). This assessment ensures that the reports align with stakeholder concerns and priorities, thereby offering transparency (Driver & ElAlfy, 2023; Mio et al., 2024) and an improved capability of detecting risks (Panfilo et al., 2025). GRI standards advocate a stakeholder-inclusive approach that compels the reporting organization to consider the broader impact of its actions and the interests of external parties.

There have been varying legislative attempts to incorporate materiality into sustainability reporting. European legislation has increasingly embraced the concept of double materiality, particularly through initiatives like the Non-Financial Reporting Directive (NFRD) and, more recently, the Corporate Sustainability Reporting Directive (CSRD) (Fiandrino et al., 2022). These Directives push companies toward more comprehensive disclosures that clarify the business's impact on the environment and society, and vice versa (Förster, 2022; see also Sect. 2.3). According to the European perspective, the updated exposure draft ESRS 2 (general disclosures) published in July 2025 states, "The undertaking shall disclose a concise description of its actual and potential, positive and negative material impacts, including how they affect or are likely to affect people or the environment, as well as its material risks and opportunities, specifying the related topics and how and where impacts, risks and opportunities are connected to its business model, its own operations and its upstream and downstream value chain" (p. 13).

Meanwhile, the US approach has traditionally focused more on financial materiality, with bodies such as the Securities and Exchange Commission (SEC) building guidelines around the importance of material information for investors. In China, there is a growing emphasis on integrating social and environmental dimensions, with a focus that is mainly tilted toward operational transparency and improving the comprehensive measurement of sustainability impacts (Shen et al., 2020). China has moved decisively from experimenting with ESG pilots to hard-wiring double materiality in its rulebook. A recent step in this direction came on December 17,

2024, when the Ministry of Finance's Corporate Sustainability Disclosure Standards—Basic Standards (Trial) formally required companies to analyze every sustainability topic through two lenses: (i) how it affects enterprise value and (ii) how corporate activities affect people, the economy, and the environment. Although the application is voluntary until 2026 and universal only by 2030, the text makes the double materiality test the future legal minimum for operations in China (Interesse, 2024). Less than 5 months earlier, the Shanghai, Shenzhen, and Beijing exchanges translated the principle into practice. Their Sustainability Report (Trial) guidelines, effective May 1, 2024, oblige companies in flagship indices to publish a 2025 report covering 21 ESG topics, each screened for both financial and impact relevance. The guidance even explains how to judge time horizons and balance probabilities against the scale of external harm by anchoring the ISSB baseline in a CSRD-oriented scheme (KPMG, 2024). Commentators see strategic positioning rather than mere copy-and-paste. UNEP FI argues that by adopting Europe's double materiality perspective while phasing implementation more gently, Beijing seeks to offer global investor comparability without overwhelming domestic SMEs. This approach could serve as a bridge between the investor-only ISSB model and the stakeholder-rich EU regime (UNEP FI, 2025). Crucially, this new architecture is not being built in a vacuum. The Ministry of Ecology and Environment's 2022 Measures on the Legal Disclosure of Enterprise Environmental Information have already forced thousands of high-impact firms to publish detailed emissions and compliance data, giving regulators a ready-made "impact materiality" spine that now plugs into the wider double-materiality approach (Huld, 2022). The result is a phased yet irreversible convergence: Chinese enterprises are first learning to quantify outward effects and then mapping how those same issues feed back into cash flows and risk. By the end of the decade, China is set to stand on level terms with international disclosure norms, as broadly discussed in the Sect. 3.2. "The comparison between different approaches."

The shift toward double materiality is increasingly seen as a necessary evolution in sustainability reporting, even though there have only been a few practical implementations (Laine, 2024). In this vein, authors like Barker and Mayer (2025) have argued that double materiality not only enhances the comprehensiveness of reports but also aligns corporate objectives with global sustainability goals, such as those outlined in the Paris Agreement and the SDGs. This approach is expected to become more prevalent as stakeholders demand greater accountability, which will create a demand for more verifiable information. Appelbaum et al. (2023) emphasized the importance of auditing practices in the realm of ESG disclosures, highlighting how "the double materiality audit" can provide assurance that organizations are addressing the impacts they both generate and endure. Robust auditing mechanisms can verify the accuracy and relevance of disclosed information, ensuring that it meets stakeholders' demands for transparency and accountability (Mezzanotte, 2023). In line with this reasoning, Turner and Weirich (2023) expanded the traditional concept of materiality to encompass environmental, social, and governance aspects, reshaping audit practices and implications, thereby enhancing the reliability of disclosures and fostering trust-based relations among investors and other stakeholders.

This element of trust between a company and its stakeholders exposes the absence of a counterpart that regularly verifies the claims made. Given the lack of audits or verification processes—particularly for issues with impacts that are difficult to measure—it is essential to adopt a stringent and reliable materiality assessment.

Of course, there are various ways that firms can implement double materiality, which can complicate the situation. In this regard, Bayne's (2022) framework usefully broke the concept down into key parts. His classification distinguishes between the reporting 'entity' (Financial vs. Sustainability Control/Influence); 'target users' (Investor vs. Wider Stakeholder Focus); 'materiality' (Financial vs. Enviro-Socio-Economic); 'boundary description' (Entity-wide vs. Topic Boundary); 'impact' (Outward vs. Inward); 'outward impact' (Direct vs. Indirect); 'time' (Historic vs. Future); 'performance' (Financial vs. Sustainability); 'value' (Entity vs. Wider Stakeholder), and 'purpose.' In doing so, he mapped the boundary dimensions of reporting regulations and guidelines (FRC, 2018; GRI, 2020; IASB, 2018–2021; IFRS Foundation, 2020–2021; IIRC, 2021; NFRD Proposal of EU Commission, 2021; SASB, 2020; TCFD, 2017) and found significant differences between the approaches (see also his annex and Sect. 3.2).

Furthermore, a recent graphic published by the Chief Sustainability Officers Network (2024) clearly visualizes the strength of the European approach, specifically EGFRAG, regarding double materiality logics. Figure 2.3 illustrates the different layers of this intervention: the standard setters, the reporting standards, and the correlated climate reporting systems.

In short, the evolution of materiality in corporate sustainability reporting reflects a broad shift in the business landscape. Driven in part by empirical findings,

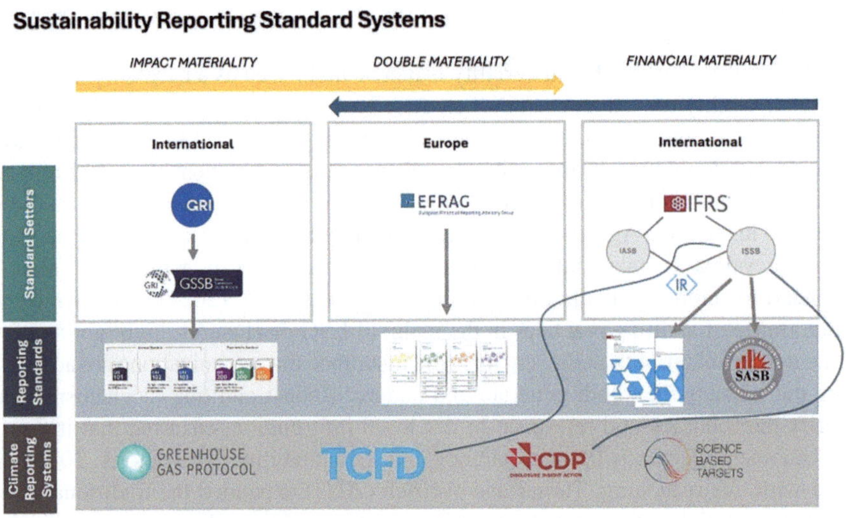

Fig. 2.3 Sustainability Reporting Standard Systems and the different approaches to materiality. Source: Chief Sustainability Officers Network (2024)

practitioners and policymakers increasingly recognize the interconnectedness of financial performance and sustainability impacts (Mio et al., 2023).

This chapter has explored how sustainability reporting and performance measurement are deeply influenced by interconnected stakeholders, value chain dynamics, and the need for robust materiality assessments. These insights highlight the importance of shared definitions, standardized frameworks, and transparent communication for ensuring accountability and improving decision-making. This evolution can be supported or limited by both regulatory changes and a deeper understanding of the impacts.

Moving forward, Chap. 3 will explore the institutional and normative context that shapes sustainability practices. It will examine how regulatory approaches—particularly within Europe—have evolved to drive or hinder sustainable development. Specific directives, such as the Non-Financial Reporting Directive (NFRD), Corporate Social Responsibility Directive (CSRD), and Corporate Sustainability Due Diligence Directive (CSDDD), illustrate how formalized norms influence organizational reporting, data collection, and overall reliability. This institutional lens will provide a clearer understanding of how external pressures and policy frameworks guide the adoption and refinement of sustainability strategies.

References

Adams, C. A. (2015). The international integrated reporting council: A call to action. *Critical Perspectives on Accounting, 27*, 23–28.

Adams, C. A. (2017). The sustainable development goals, integrated thinking and the integrated report. *Integrated Reporting (IR)*, 1–52.

Adams, C. A., & Frost, G. R. (2008). Integrating sustainability reporting into management practices. *Accounting Forum, 32*(4), 288–302.

Adams, C. A., Hill, W. Y., & Roberts, C. B. (1998). Corporate social reporting practices in Western Europe: Legitimating corporate behaviour? *The British Accounting Review, 30*(1), 1–21.

Adams, C. A., & McNicholas, P. (2007). Making a difference: Sustainability reporting, accountability and organisational change. *Accounting, Auditing & Accountability Journal, 20*(3), 382–402.

Ahi, P., & Searcy, C. (2015). An analysis of metrics used to measure performance in green and sustainable supply chains. *Journal of Cleaner Production, 86*, 360–377.

Aldhizer, G. R., III, Martin, D. R., & Cotter, J. F. (2009). Do markets react to required and voluntary disclosures associated with auditor realignments? *Advances in Accounting, 25*(1), 1–12. https://doi.org/10.1016/j.adiac.2009.02.002

Allais, R., Roucoules, L., & Reyes, T. (2017). Governance maturity grid: A transition method for integrating sustainability into companies? *Journal of Cleaner Production, 140*, 213–226.

Al-Tuwaijri, S. A., Christensen, T. E., & Hughes, K. E., II. (2004). The relations among environmental disclosure, environmental performance, and economic performance: A simultaneous equations approach. *Accounting, Organizations and Society, 29*, 447–471.

Antonini, C., Beck, C., & Larrinaga, C. (2020). Subpolitics and sustainability reporting boundaries. The case of working conditions in global supply chains. *Accounting, Auditing & Accountability Journal, 33*(7), 1535–1567.

Appannan, J. S., Mohd Said, R., Ong, T. S., & Senik, R. (2023). Promoting sustainable development through strategies, environmental management accounting and environmental performance. *Business Strategy and the Environment, 32*(4), 1914–1930.

Appelbaum, D., Duan, H. K., Hu, H., & Sun, T. (2023). *The double materiality audit: Assurance of ESG disclosure*. Available at SSRN 4367032.

Asif, M. S., Lau, H., Nakandala, D., Fan, Y., & Hurriyet, H. (2022). Case study research of green life cycle model for the evaluation and reduction of scope 3 emissions in food supply chains. *Corporate Social Responsibility and Environmental Management, 29*(4), 1050–1066. https://doi.org/10.1002/csr.2253

Aureli, S., Del Baldo, M., Lombardi, R., & Nappo, F. (2020). Nonfinancial reporting regulation and challenges in sustainability disclosure and corporate governance practices. *Business Strategy and the Environment, 29*(6), 2392–2403.

Azzone, G., & Noci, G. (1998). Identifying effective PMSs for the deployment of "green" manufacturing strategies. *International Journal of Operations & Production Management, 18*(4), 308–335.

Azzone, G., Noci, G., Manzini, R., Welford, R., & Young, C. W. (1996). Defining environmental performance indicators: An integrated framework. *Business Strategy and the Environment, 5*(2), 69–80. https://doi.org/10.1002/(sici)1099-0836(199606)5:2<69::aid-bse48>3.0.co;2-w

Baah, C., Opoku-Agyeman, D., Acquah, I. S. K., Agyabeng-Mensah, Y., Afum, E., Faibil, D., & Abdoulaye, F. A. M. (2021). Examining the correlations between stakeholder pressures, green production practices, firm reputation, and environmental and financial performance: Evidence from manufacturing SMEs. *Sustainable Production and Consumption, 27*, 100–114.

Baldini, M., Maso, L. D., Liberatore, G., Mazzi, F., & Terzani, S. (2018). Role of country-and firm-level determinants in environmental, social, and governance disclosure. *Journal of Business Ethics, 150*, 79–98.

Bansal, P., & Clelland, I. (2004). Talking trash: Legitimacy, impression management, and unsystematic risk in the context of the natural environment. *Academy of Management Journal, 47*(1), 93–103.

Barker, R., & Mayer, C. (2025). Seeing double corporate reporting through the materiality lenses of both investors and nature. *Accounting Forum, 49*(2), 259–289.

Baumgartner, R. J., & Ebner, D. (2010). Corporate sustainability strategies: Sustainability profiles and maturity levels. *Sustainable Development, 18*(2), 76–89.

Bayne, L. (2022). Understanding reporting boundaries in annual reports: A conceptual framework. *Accounting, Auditing & Accountability Journal, 35*(5), 1316–1348.

Bebbington, J., Russell, S., & Thomson, I. (2017). Accounting and sustainable development: Reflections and propositions. *Critical Perspectives on Accounting, 48*, 21–34.

Benston, G. J. (1982). Accounting and corporate accountability. *Accounting, Organizations and Society, 7*(2), 87–105.

Beske-Janssen, P., Johnson, M. P., & Schaltegger, S. (2015). 20 years of performance measurement in sustainable supply chain management–what has been achieved? *Supply Chain Management: An international Journal, 20*(6), 664–680.

Boiral, O. (2016). Accounting for the unaccountable: Biodiversity reporting and impression management. *Journal of Business Ethics, 135*(4), 751–768.

Boiral, O., & Henri, J. F. (2017). Is sustainability performance comparable? A study of GRI reports of mining organizations. *Business & Society, 56*(2), 283–317.

Boissinot, J., Goulard, S., Salin, M., Svartzman, R., & Weber, P. F. (2022). *Aligning financial and monetary policies with the concept of double materiality: Rationales, proposals and challenges*. http://eprints.lse.ac.uk/id/eprint/115539

Boone, T., Jayaraman, V., & Ganeshan, R. (2012). *Sustainable supply chains* (pp. 1–8). Springer.

Brown, J., & Dillard, J. (2014). Integrated reporting: On the need for broadening out and opening up. *Accounting, Auditing & Accountability Journal, 27*(7), 1120–1156.

Bunclark, L., & Barcellos-Paula, L. (2021). Sustainability reporting for sustainable supply chain management in Peru. *Sustainable Production and Consumption, 27*, 1458–1472.

Burritt, R., & Schaltegger, S. (2014). Accounting towards sustainability in production and supply chains. *The British Accounting Review, 46*(4), 327–343.

Busco, C., & Sofra, E. (2021). The evolution of sustainability reporting: Integrated reporting and sustainable development challenges. In *Corporate sustainability in practice: A guide for strategy development and implementation* (pp. 191–206). Springer Nature.

Caccialanza, A., & Torelli, R. (2024). The role of trade associations in promoting corporate sustainability transition and reporting: A case study in the food supply chain. *Corporate Social Responsibility and Environmental Management, 31*(5), 4469–4486.

Cao, S. S., Fang, V. W., & Lei, L. (2021). Negative peer disclosure. *Journal of Financial Economics, 140*(3), 815–837. https://doi.org/10.1016/j.jfineco.2021.02.007

Chen, L., Feldmann, A., & Tang, O. (2015). The relationship between disclosures of corporate social performance and financial performance: Evidences from GRI reports in manufacturing industry. *International Journal of Production Economics, 170*, 445–456.

Chiarini, A. (2017). Environmental policies for evaluating suppliers' performance based on GRI indicators. *Business Strategy and the Environment, 26*(1), 98–111.

Chief Sustainability Officers Network. (2024). *Post published by Philipp Schröder, Ph.D.* https://www.linkedin.com/feed/update/urn:li:activity:7284469627521253376?utm_source=share&utm_medium=member_desktop

Cho, C. H., & Patten, D. M. (2007). The role of environmental disclosures as tools of legitimacy: A research note. *Accounting, Organizations and Society, 32*(7–8), 639–647.

Cooper, S., & Michelon, G. (2022). Conceptions of materiality in sustainability reporting frameworks: Commonalities, differences and possibilities. In *Handbook of accounting and sustainability* (pp. 44–66). Edward Elgar Publishing.

Cooper, S. M., & Owen, D. L. (2007). Corporate social reporting and stakeholder accountability: The missing link. *Accounting, Organizations and Society, 32*(7–8), 649–667.

Cormier, D., Gordon, I. M., & Magnan, M. (2004). Corporate environmental disclosure: Contrasting management's perceptions with reality. *Journal of Business Ethics, 49*, 143–165.

Costa, E., & Pesci, C. (2016). Social impact measurement: Why do stakeholders matter? *Sustainability Accounting, Management and Policy Journal, 7*(1), 99–124.

Dal Maso, L., Liberatore, G., & Mazzi, F. (2017). Value relevance of stakeholder engagement: The influence of national culture. *Corporate Social Responsibility and Environmental Management, 24*(1), 44–56.

Dal Maso, L., Lobo, G. J., Mazzi, F., & Paugam, L. (2020). Implications of the joint provision of CSR assurance and financial audit for auditors' assessment of going-concern risk. *Contemporary Accounting Research, 37*(2), 1248–1289.

Dal Maso, L., Mazzi, F., Soscia, M., & Terzani, S. (2018). The moderating role of stakeholder management and societal characteristics in the relationship between corporate environmental and financial performance. *Journal of Environmental Management, 218*, 322–332.

Dietz, T., Grabs, J., & Chong, A. E. (2021). Mainstreamed voluntary sustainability standards and their effectiveness: Evidence from the Honduran coffee sector. *Regulation & Governance, 15*(2), 333–355.

Dodsworth, C., Bourne, E., Schlich, B. L., & Kooroshy, J. (2023). Financial, double, or dynamic? Theories of ESG materiality and practitioner approaches. *Journal of Impact & ESG Investing, 4*(2), 109.

Dragomir, V. D., Dumitru, M., Chersan, I. C., Gorgan, C., & Păunescu, M. (2025). Double materiality disclosure as an emerging practice: The assessment process, impacts, risks, and opportunities. *Accounting in Europe, 22*(1), 103–140.

Dragu, I. (2019). The evolution of corporate sustainability and corporate social responsibility towards the common goal of integrated reporting. In *Integrated reporting: Antecedents and perspectives for organizations and stakeholders* (pp. 77–93). Springer.

Driver, T. R., & ElAlfy, A. (2023). Double materiality: Why does it matter for sustainability reporting? In *The Routledge handbook of green finance* (pp. 41–56). Routledge.

Edgley, C., Jones, M. J., & Atkins, J. (2015). The adoption of the materiality concept in social and environmental reporting assurance: A field study approach. *The British Accounting Review, 47*(1), 1–18.

Elkington, J. (1998). Accounting for the triple bottom line. *Measuring Business Excellence, 2*, 18–22.

Epstein, M. J. (2007). *Measuring corporate environmental performance: Best practices for costing and managing an effective environmental strategy,* 16, 6th ed., 389–403.

Fekrat, M. A., Inclan, C., & Petroni, D. (1996). Corporate environmental disclosures: Competitive disclosure hypothesis using 1991 annual report data. *International Journal of Accounting, 31*(2), 175–195. https://doi.org/10.1016/S0020-7063(96)90003-5

Fiandrino, S., Gromis di Trana, M., Tonelli, A., & Lucchese, A. (2022). The multi-faceted dimensions for the disclosure quality of non-financial information in revising directive 2014/95/EU. *Journal of Applied Accounting Research, 23*(1), 274–300.

Fontoura, P., & Coelho, A. (2022). How to boost green innovation and performance through collaboration in the supply chain: Insights into a more sustainable economy. *Journal of Cleaner Production, 359*, 132005.

Foran, B., Lenzen, M., Dey, C., & Bilek, M. (2005). Integrating sustainable chain management with triple bottom line accounting. *Ecological Economics, 52*(2), 143–157.

Förster, P. (2022). The double materiality principle (article 19a NFRD) as proposed by the corporate sustainability reporting directive: An effective concept to tackle green washing? In *European yearbook of international economic law 2022* (pp. 345–364). Springer International Publishing.

Fritz, M. M., Schöggl, J. P., & Baumgartner, R. J. (2017). Selected sustainability aspects for supply chain data exchange: Towards a supply chain-wide sustainability assessment. *Journal of Cleaner Production, 141*, 587–607.

Gagalyuk, T., Chatalovaa, L., Kalyuzhnyy, O., & Ostapchukc, I. (2021). Broadening the scope of instrumental motivations for CSR disclosure: An illustration for Agroholdings in transition economies. *International Food and Agribusiness Management Review, 24*(4), 717–737. https://doi.org/10.22434/IFAMR2020.0210

Galli, D., Torelli, R., & Caccialanza, A. (2023). Sustainability performance and sustainability reporting in SMEs: A love affair or a fight? *Journal of Management & Organization, 30*, 1–26.

Göbbels, M., & Jonker, J. (2003). AA1000 and SA8000 compared: A systematic comparison of contemporary accountability standards. *Managerial Auditing Journal, 18*(1), 54–58.

Goharshenasan, A., Aboumasoudi, A. S., Shahin, A., & Ansari, A. (2022). Identifying and classifying sustainable supply chain performance indicators: A GRI-based multivariate analysis. *International Journal of Industrial and Systems Engineering, 41*(1), 41–70.

Grabs, J., & Garrett, R. D. (2023). Goal-based private sustainability governance and its paradoxes in the Indonesian palm oil sector. *Journal of Business Ethics, 188*(3), 467–507.

Gray, R. (2001). Thirty years of social accounting, reporting and auditing: What (if anything) have we learnt? *Business Ethics: A European Review, 10*(1), 9–15.

Gray, R. (2013). Back to basics: What do we mean by environmental (and social) accounting and what is it for? A reaction to Thornton. *Critical Perspectives on Accounting, 24*(6), 459–468.

Gray, R., Owen, D., & Adams, C. (1996). *Accounting and accountability: Changes and challenges in corporate social and environmental reporting*. Prentice-Hall.

Gualandris, J., Klassen, R. D., Vachon, S., & Kalchschmidt, M. (2015). Sustainable evaluation and verification in supply chains: Aligning and leveraging accountability to stakeholders. *Journal of Operations Management, 38*, 1–13.

Hahn, R., & Kühnen, M. (2013). Determinants of sustainability reporting: A review of results, trends, theory, and opportunities in an expanding field of research. *Journal of Cleaner Production, 59*, 5–21.

Hettler, M., & Graf-Vlachy, L. (2024). Corporate scope 3 carbon emission reporting as an enabler of supply chain decarbonization: A systematic review and comprehensive research agenda. *Business Strategy and the Environment, 33*(2), 263–282. https://doi.org/10.1002/bse.3486

Holder-Webb, L., Cohen, J. R., Nath, L., & Wood, D. (2009). The supply of corporate social responsibility disclosures among US firms. *Journal of Business Ethics, 84*(4), 497–527.

Hossain, A., Bose, S., & Shamsuddin, A. (2023). Diffusion of integrated reporting, insights and potential avenues for future research. *Accounting & Finance, 63*(2), 2503–2555.

Huld, A. (2022, February 23). *China ESG reporting—new measures on disclosure of enterprise environmental information*. China Briefing. https://www.china-briefing.com/news/china-esg-reporting-disclosing-enterprise-environmental-information/

IFRS Foundation. (2021). *exposure draft proposed targeted amendments to the IFRS foundation constitution to accommodate an international sustainability standards board to Set IFRS sustainsability standards*. IFRS Foundation.

IFRS Foundation Consultation Paper (IFRS Foundation, 2020) and ED (IFRS Foundation, 2021)—IFRS Foundation. (2020). *Consultation paper on sustainability reporting*. IFRS Foundation.

IFRS-CF (IASB, 2018)—International Accounting Standards Board. (2018). *Conceptual framework for financial reporting*. International Financial Reporting Standards (IFRS).

IFRS-MC-ED (IASB, 2021)—International Accounting Standards Board. (2021). *Practice statement exposure draft ED/2021/6 management commentary*. International Financial Reporting Standards (IFRS).

Interesse, G. (2024, December 23). *China unveils its first set of basic standards for corporate sustainability (ESG) disclosure*. China Briefing. https://www.china-briefing.com/news/china-unveils-basic-standards-for-corporate-sustainability-esg-disclosure/

Jørgensen, S., Mjøs, A., & Pedersen, L. J. T. (2022). Sustainability reporting and approaches to materiality: Tensions and potential resolutions. *Sustainability Accounting, Management and Policy Journal, 13*(2), 341–361.

Karatzoudi, K., & Aven, T. (2024). Principles and guidance on climate risk disclosure for businesses. *International Journal of Business Continuity and Risk Management, 14*(1), 14–29. https://doi.org/10.1504/IJBCRM.2024.137237

KPMG China. (2024, April 12). *China stock exchanges finalised mandatory sustainability reporting requirements for larger listed entities*. https://assets.kpmg.com/content/dam/kpmg/cn/pdf/en/2024/04/china-stock-exchanges-finalised-mandatory-sustainability-reporting-requirements-for-larger-listed-entities.pdf

Kundeliene, K., & Stepanauskaite, A. (2018). Information disclosure and its determinants on Lithuanian companies' websites: A quantitative approach. *Engineering Economics, 29*(4), 455–467. https://doi.org/10.5755/j01.ee.29.4.17271

Lai, A., & Stacchezzini, R. (2021). Organisational and professional challenges amid the evolution of sustainability reporting: A theoretical framework and an agenda for future research. *Meditari Accountancy Research, 29*(3), 405–429.

Laine, J. (2024). *Double materiality assessment: Early practices and process disclosures*.

Lamming, R., Caldwell, N., Phillips, W., & Harrison, D. (2005). Sharing sensitive information in supply relationships: The flaws in one-way open-book negotiation and the need for transparency. *European Management Journal, 23*(5), 554–563.

Lee, K. H., & Kim, J. W. (2009). Current status of CSR in the realm of supply management: The case of the Korean electronics industry. *Supply Chain Management: An International Journal, 14*(2), 138–148.

Long, Y., Feng, T., Fan, Y., & Liu, L. (2023). Adopting blockchain technology to enhance green supply chain integration: The moderating role of organizational culture. *Business Strategy and the Environment, 32*(6), 3326–3343.

Machado, M. C., Correa, V. S., de Queiroz, M. M., & Costa, G. C. (2023). Can global reporting initiative reports reveal companies' green supply chain management practices? *Journal of Cleaner Production, 383*, 135554.

Massari, G. F., & Giannoccaro, I. (2023). Adopting GRI standards for the circular economy strategies disclosure: The case of Italy. *Sustainability Accounting, Management and Policy Journal, 14*(4), 660–694.

Mezzanotte, F. E. (2023). *Examining the reasons for impact materiality in EU corporate sustainability reporting*. European Business Law Review (forthcoming).
Mio, C., Agostini, M., & Scarpa, F. (2024). *Sustainability reporting: Conception: International approaches and double materiality in action*. Springer Nature.
Mio, C., Fasan, M., & Scarpa, F. (2023). Materiality investor perspectives on utilities' ESG performance. An empirical analysis of ESG factors and cost of equity. *Utilities Policy, 82*, 101555.
Moggi, S. (2023). Sustainability reporting, universities and global reporting initiative applicability: A still open issue. *Sustainability Accounting, Management and Policy Journal, 14*(4), 699–742.
Moggi, S., Leardini, C., & Rossi, G. (2015). Mandatory or not mandatory reporting? Insights from Italian bank foundations. *International Journal of Public Administration, 38*(10), 734–742.
Okada, Y., Naito, Y., & Nagaoka, S. (2018). Making the patent scope consistent with the invention: Evidence from Japan. *Journal of Economics and Management Strategy, 27*(3), 607–625. https://doi.org/10.1111/jems.12266
Oktorina, M., Siregar, S. V., Adhariani, D., & Mita, A. F. (2022). The diffusion and adoption of integrated reporting: A cross-country analysis on the determinants. *Meditari Accountancy Research, 30*(1), 39–73.
Panfilo, S., Scarpa, F., & Canestraro, N. (2025). Reporting for change: Does the adoption of double materiality influence ESG risk management? *Management Decision*.
Panjwani, A., Melin, L., & Mercereau, B. (2023). Do scope 3 carbon emissions impact firms' cost of debt? *Journal of Portfolio Management, 50*(2), 175–198. https://doi.org/10.3905/jpm.2023.1.548
Parmigiani, A., Klassen, R. D., & Russo, M. V. (2011). Efficiency meets accountability: Performance implications of supply chain configuration, control, and capabilities. *Journal of Operations Management, 29*(3), 212–223.
Perego, P., Kennedy, S., & Whiteman, G. (2016). A lot of icing but little cake? Taking integrated reporting forward. *Journal of Cleaner Production, 136*, 53–64.
Pérez-López, D., Moreno-Romero, A., & Barkemeyer, R. (2015). Exploring the relationship between sustainability reporting and sustainability management practices. *Business Strategy and the Environment, 24*(8), 720–734.
Pesci, C., Girardi, A., & Moggi, S. (2025). Social distance, accounting and accountability in non-profit organisations. In *Accounting, accountability and crisis management* (pp. 150–165). Routledge.
Potrich, L., Cortimiglia, M. N., & de Medeiros, J. F. (2019). A systematic literature review on firm-level proactive environmental management. *Journal of Environmental Management, 243*, 273–286.
Qian, W., & Schaltegger, S. (2017). Revisiting carbon disclosure and performance: Legitimacy and management views. *British Accounting Review, 49*(4), 365–379. https://doi.org/10.1016/j.bar.2017.05.005
Rasche, A., & Esser, D. E. (2006). From stakeholder management to stakeholder accountability: Applying Habermasian discourse ethics to accountability research. *Journal of Business Ethics, 65*, 251–267.
Rhianon Edgley, C., Jones, M. J., & Solomon, J. F. (2010). Stakeholder inclusivity in social and environmental report assurance. *Accounting, Auditing & Accountability Journal, 23*(4), 532–557.
Roberts, L., & Gehrke, T. (1996). Linkages between best practice in business and good environmental performance by companies. *Journal of Cleaner Production, 4*(3–4), 189–202.
Robertson, F. A., & Samy, M. (2015). Factors affecting the diffusion of integrated reporting–a UK FTSE 100 perspective. *Sustainability Accounting, Management and Policy Journal, 6*(2), 190–223.
Setia, N., Abhayawansa, S., Joshi, M., & Wasantha Pathiranage, N. (2024). Shifting perspectives: Unveiling the dual nature of sustainability materiality in integrated reports. *Meditari Accountancy Research, 32*(4), 1291–1323.

Shearer, T. (2002). Ethics and accountability: From the for-itself to the for-the-other. *Accounting, Organizations and Society, 27*(6), 541–573.

Shen, H., Ng, A. W., Zhang, J., & Wang, L. (2020). Sustainability accounting, management and policy in China: Recent developments and future avenues. *Sustainability Accounting, Management and Policy Journal, 11*(5), 825–839.

Shukla, G. P., & Adil, G. K. (2021). A four-stage maturity model of green manufacturing orientation with an illustrative case study. *Sustainable Production and Consumption, 26*, 971–987.

Sinclair, A. (1995). The chameleon of accountability: Forms and discourses. *Accounting, Organizations and Society, 20*(2/3), 219–237.

Stranieri, S., Orsi, L., Banterle, A., & Ricci, E. C. (2019). Sustainable development and supply chain coordination: The impact of corporate social responsibility rules in the European Union food industry. *Corporate Social Responsibility and Environmental Management, 26*(2), 481–491.

Stroehle, J. C., Soonawalla, K., & Metzner, M. (2022). Through the looking glass: Tying performance and materiality to corporate purpose. *Journal of the British Academy, 10*(5), 87–123.

Stubbs, W., & Higgins, C. (2014). Integrated reporting and internal mechanisms of change. *Accounting, Auditing & Accountability Journal, 27*(7), 1068–1089.

Suwaidan, M. S., Al-Omari, A. M., & Haddad, R. H. (2004). Social responsibility disclosure and corporate characteristics: The case of Jordanian industrial companies. *International Journal of Accounting, Auditing and Performance Evaluation, 1*(4), 432–447. https://doi.org/10.1504/ijaape.2004.006355

Turner, L., & Weirich, T. R. (2023). Expanding the concept of materiality to environmental, social, and governance: Audit issues and implications. *Current Issues in Auditing, 17*(1), A50–A58.

United Nations Environment Programme Finance Initiative. (2025, January 7). *China embarks on a journey of ESG disclosure: 2024 progress and focus for 2025.* https://www.unepfi.org/industries/banking/china-embarks-on-a-journey-of-esg-disclosure/

Winn, M. L., & Angell, L. C. (2000). Towards a process model of corporate greening. *Organizsation Studies, 21*(6), 1119–1147.

Woodward, D. G., Edwards, P., & Birkin, F. (1996). Organizational legitimacy and stakeholder information provision. *British Journal of Management, 7*(4), 329–347.

Yasar, A., & Yalçın, N. (2023). Voluntary disclosure of scope 3 greenhouse gas emissions and earnings management: Evidence from UK companies. *Cogent Business and Management, 10*(3), 2275849. https://doi.org/10.1080/23311975.2023.2275849

Chapter 3
The Role of Institutions and Norms

3.1 Institutional and Normative Drivers and Barriers

This subsection introduces the limits and boundary conditions for the status quo of sustainable supply chain production models. This section discusses several contributions from the literature on mitigating the relevant barriers and enhancing the implementation and diffusion of sustainability (see the drivers and barriers in Fig. 5.1). To this end, it is useful to envision sustainable supply chain management as a pyramid: According to Olsen and Boxenbaum (2009), the bottom of the pyramid can be defined as "the creation of new profit-seeking market opportunities for low-income segments in the developing world with the simultaneous goal of contributing to the resolution of significant societal problems in these regions" (p. 102). In other words, placing SSCM at the base of the pyramid underscores its innovative impact and contribution to the creation of shared value (Khalid et al., 2015; Khalid & Seuring, 2019). It is important to emphasize that sustainable managerial practices should be adopted as systemically as possible at the supply chain level and integrated at the firm level (Morali & Searcy, 2013). This has a dual effect: On the one hand, it promotes the dissemination of best practices among a broader network of actors rather than isolated entities; on the other hand, it increases the effect size of the positive impacts generated or reduces the negative ones (Martínez-Ferrero & García-Sánchez, 2017).

Scholars such as Diabat et al. (2014) reviewed and discussed several enablers of sustainability integration, such as the adoption of safety standards and green practices, community economic welfare, health and safety issues, and employment stability. That said, the literature has identified various barriers to the implementation of sustainable supply chain practices, both internal and external to the company. On the external factors side, Govindan and Bouzon (2018) emphasized how the lack of facilities (such as deficient industrial infrastructure) in the institutional context can inhibit. Regarding internal factors, they

A. Caccialanza, *Sustainability and Accountability of Supply Chains*, SIDREA Series in Accounting and Business Administration,
https://doi.org/10.1007/978-3-032-16879-5_3

underlined the issues of employees' skill level (whether declared or acted upon) regarding the change in processes and management of (incremental) operational complexity.

These dynamics prompt a reflection on the current state of practical implementations and the limitations of the approaches pursued thus far. In fact, the application of the current approach has been significantly limited by the failure to spread this principle. In a vicious and circular manner, the persistence of certain incorrect and unsustainable practices would inevitably backfire on operators who, despite the risk of being burdened with higher-cost structures, decide to transition their production model for waste disposal (Troisi et al., 2024). Transition risks are among the first factors that an operator evaluates in the decision-making process before the moment of change (for the production model) (Fan & Stevenson, 2018). The second problem involves the lack of uniformity in defining what is sustainable and which method(s) should be used to assess performance. Practitioners lack a clear international standard or framework for not only measurements but also verifiability (i.e., checking claims about the final product's sustainability) (Das, 2017). As has happened in Western and mature markets, product sustainability claims may emerge that are not verified by a third party and thus not strictly certified (Pedersen & Andersen, 2023). This issue is exacerbated by the proliferation of non-interoperable and non-comparable platforms and systems (Kazancoglu et al., 2023). If correctly implemented in current systems, however, new technology might enable a transparent and integrated approach to sustainability transition. Of course, the utility of such technology depends on governance factors, which represent another major obstacle. Indeed, a lack of effective leaders (of change) and an emphasis on cost reduction can hinder the effective implementation of these strategies and the spread of the abovementioned sustainability practices (Formentini & Taticchi, 2016). In this sense, legislators play a fundamental role in implementing measures and incentives (particularly related to taxes) that can facilitate the transition of production models (Camilleri, 2020; Maas & Vermeulen, 2015; Scarpa et al., 2025). Of course, there are pitfalls here, too: For instance, ambitious national strategic plans may not find international recognition, thus limiting their feasibility or scope (Pickering et al., 2020). Conversely, measures that are too restrictive in terms of environmental and social protection can trigger crises in existing production systems, leading to the relocation of the most sensitive and impactful production processes and maybe a loss in market competitiveness (John et al., 2016). Furthermore, forms of inertia in traditional and mature production systems may delay the adoption of innovations (Abbasi & Nilsson, 2012), as can a lack of specific competencies (Beske-Janssen et al., 2023). Against this background, there is a need to not only involve more disciplines in the conversation but also find and legitimize international actors and regulatory sources that can overcome the current limitations of disjointed and competing national legislations (Camilleri, 2020; Khalid & Seuring, 2019). In this context of fragmented and (frequently) incomparable results, new European regulations represent a promising attempt to alter the status quo.

3.2 The Comparison Between Different Approaches

Sustainability reporting has become a key tool for organizations aiming to demonstrate accountability in their practices while adapting their strategies to new market realities (Davis & DeWitt, 2024; Tate et al., 2010; Wieland, 2021). That said, the regulatory frameworks that govern the drafting of sustainability reports are region-specific, with each featuring unique priorities and challenges that influence corporate reporting behaviors. In this section, we focus on three contexts that represent large parts of the global economy: Europe, China, and the United States.

Focusing on the regulatory approaches in these three regions can be considered a valuable research strategy since they dominate both the global economy and import-export trades, thus creating powerful channels through which sustainability-reporting rules generate cross-border spillovers. In fact, together, the EU (14.1% of world GDP at purchasing-power parity), the United States (14.8%), and China (19.7%) account for roughly 48% of global output in PPP terms and about 61% in nominal terms (International Monetary Fund—IMF, 2025). When any one of them tightens disclosure standards—as the EU has done with the CSRD's double-materiality test—the regulatory shock automatically reaches nearly half of the world economy. The intensity of trade between the three blocks amplifies this effect. The EU ships 20.6% of its extra-EU exports to the United States and buys 20.1% of its extra-EU imports from China (Eurostat, 2025). In 2024 alone, the United States exchanged USD 976 billion in goods with the EU and USD 583 billion with China, running sizeable deficits with both partners (U.S. Census Bureau, 2025a, 2025b). These multi-trillion-dollar supply chains mean that European buyers will demand CSRD-compliant ESG data from American and Chinese suppliers, transmitting the EU rulebook well beyond its borders. A comparative lens is therefore indispensable for assessing externalities: the EU's broad double-materiality regime, the United States' investor-focused ISSB convergence, and China's hybrid model (now explicitly embedding double materiality as mentioned in the dedicated Sect. 2.4, Double Materiality) interact in ways that can either raise or lower compliance costs for firms operating across the three markets (Wall Street Journal, 2025). Analyzing them together clarifies where regulatory frictions may emerge, where interoperability is already taking shape, and how global companies can design reporting systems that satisfy all three giants without duplicating effort. When we frame our comparative analysis around the European Union, China, and the United States, we are not claiming that these three jurisdictions exhaust the universe of meaningful ESG-reporting regimes; rather, they anchor the two most consequential comparative axes for assessing how EU sustainability legislation (notably the CSRD and its double-materiality mandate) generates regulatory externalities across global value chains. Their outsized weight in global output and finance means that changes in any one of the three quickly propagate through multinational reporting systems and investor due diligence workflows (IMF, 2025). The EU–U.S. axis matters because U.S. practice remains rooted in investor-centric disclosure (historically SEC materiality; now moving toward ISSB alignment), offering a high-contrast foil to Europe's broader stakeholder scope. Firms straddling the Atlantic must reconcile financial materiality baselines with EU-required impact metrics—a reconciliation that drives cost, data architecture, and assurance strategy decisions (Wall Street Journal, 2025). The EU–China axis is equally strategic: China's emerging hybrid (embedding double materiality in national guidelines while phasing implementation) intersects with dense EU–China trade and supplier relationships, so European disclosure rules can cascade upstream into Chinese manufacturing networks, and, conversely, Chinese data requirements can reshape what European lead firms can reliably collect from their tiers of suppliers (Eurostat, 2025; U.S. Census Bureau, 2025b). Because the United States is also the EU's largest single export destination and a major import source—and because U.S. trade with both the EU and China is measured in the high hundreds of billions of dollars—these triangular flows provide a powerful conduit through which any tightening of EU sustainability-reporting obligations can influence disclosure behaviors and data quality requirements well beyond Europe's borders (Eurostat, 2025; U.S. Census Bureau, 2025a, 2025b). Sect. 5.3 of the volume digs deeper into how these two comparison points help isolate supply chain-mediated impacts—our central research lens—but we acknowledge that further comparative studies could extend the map: for example, exploring regulatory spillovers into large emerging markets (India, ASEAN), resource-exporting jurisdictions, or sector-specific regimes; examining how soft-law frameworks (ISSB, GRI) and bilateral trade instruments interact with statutory reporting mandates; and testing whether interoperability toolkits can reduce duplicative compliance burdens outside the big three blocks. These topics lie beyond the direct scope of this supply chain impact study but merit future research (Wall Street Journal, 2025).

In Europe, sustainability reporting has evolved considerably thanks to recent developments in regulatory frameworks, such as the EU Taxonomy and the guidelines issued by the European Financial Reporting Advisory Group (EFRAG). This regulatory framework is designed to facilitate consistent and comparable disclosures, driving organizations toward more robust sustainability practices (Garcia-Torea et al., 2024). This development reflects the proliferation of corporate sustainability reporting in Europe, arising from the increasing institutionalization of sustainability orientation within corporate strategies (Dinh et al., 2023). Through the directives, particularly through the directives enhancing the EFRAG's reporting standards, the EU has focused on integrated and comprehensive disclosures that cover environmental, social, and governance impacts (EFRAG, 2023b). The EU needs more thorough and transparent reporting to achieve the disclosure quality it wants from companies (e.g., González et al., 2018). One way to pressure companies to increase the quality of reporting could be the promotion of auditing processes for sustainability reporting. In line with this reasoning, Simoni et al. (2020) discussed the impact of social, environmental, and institutional factors on the assurance practices of sustainability reports in Europe. Their findings suggest that assurance practices are significantly influenced by the (European) regulatory environment, which requires more thorough and transparent reporting.

China's approach to sustainability reporting, by contrast, has been marked by a gradual increase in regulatory oversight and corporate responsibility. That said, He and Loftus (2014) provided evidence that environmental reporting in China may not always reflect the actual environmental performance. This discrepancy highlights the need for stronger enforcement mechanisms and more rigorous standards in Chinese legislation (He & Loftus, 2014). On this note, Dong et al. (2023) recently analyzed the quality of sustainability reporting within the Chinese financial sector. The authors found that the Chinese government has been implementing policies that encourage better sustainability practices, but the effectiveness of these policies is often contingent upon the regulatory environment and internal governance of the reporting entities (Dong et al., 2023). Thus, there is still considerable room for greater transparency and accountability. China's new regulation enhances both transparency and accountability by pushing firms to move fast, think holistically, and back their promises with verifiable data. First, the Sustainability Report (Trial) guidelines of the three mainland exchanges impose a hard deadline: every constituent of the SSE 180, STAR 50, SZSE 100, ChiNext, and Beijing market's large-cap index must publish a 2025 sustainability report based on a double-materiality assessment. The one-year lead time forces companies to upgrade their internal controls and disclosure pipelines almost immediately, closing the information gap that historically separated Chinese issuers from their global peers (KPMG China, 2024). Meeting that deadline without drowning in paperwork makes a rigorous gap analysis indispensable. Because the exchanges' guidelines were drafted to "promote interoperability" with the MOF's forthcoming Corporate Sustainability Disclosure Standards (CSDS), IFRS-ISSB, GRI, and CSRD, companies that map overlaps early can avoid duplicate data-gathering and strike a clearer line between investor-relevant metrics and broader impact indicators (KPMG China, 2024). At the same time, the Ministry of Ecology and Environment's 2022 Measures on the Legal Disclosure of Enterprise Environmental Information and the expanding national ETS sharpen governance and data-system expectations: Firms must install auditable ledgers, standardize terminology, and upload annual environmental returns to a public portal—non-compliance triggers fines, license revocations, or forced production cuts. By tying ESG narratives to legally validated data sets, the regime limits greenwashing and raises the reputational cost of opaque reporting (Huld, 2022), while making firms responsible for the managerial practices of their suppliers (Amaeshi et al., 2008). Finally, China's central state-owned enterprises (SOEs) act as trailblazers. SASAC's May 2022 Work Plan ordered listed SOEs to deliver full ESG reports and strengthen board-level ESG oversight, creating peer pressure for private issuers and providing real-world templates that smaller companies can emulate. The Plan's emphasis on "full coverage" by 2023 shifts ESG from marketing rhetoric to mandatory practice, anchoring accountability at the very top of China's corporate hierarchy (GoldenBee Consulting, 2023). Collectively, these measures lock companies into an accelerated learning curve (i.e., tight timelines compel action, cross-standard alignment prevents disclosure fatigue, robust data rules ensure credibility, and SOE leadership sets the cultural tone), thereby weaving transparency and accountability deep into China's corporate fabric.

Meanwhile, the United States has a more decentralized approach to sustainability reporting, often driven by market forces rather than stringent governmental regulations. Jensen and Berg (2012) described how institutional factors play a significant role in determining whether organizations opt for traditional sustainability reporting or integrated reporting. The U.S. market has historically tended to favor a voluntary approach, with companies often adopting sustainability practices as a response to investor demands rather than regulatory mandates (Jensen & Berg, 2012). Ioannou and Serafeim (2017) discussed the consequences of mandatory corporate sustainability reporting, indicating that such requirements (although not prevalent in the U.S.) could lead to improvements in both corporate reputation and operational performance. Of course, firms should be prevented from overstating their real impact (e.g., at the labor or environmental level) for the sake of improving their own reputations (e.g., as an example of this issue, see Costa & Agostini (2016) for an analysis of the Italian context).

When we broaden the focus from individual firms to supply chains, the implementation of sustainability measures varies significantly. Wu et al. (2013) performed a comparative analysis of supply chain practices among Chinese firms and foreign-owned multinationals operating in China. They found that foreign-owned companies were more often leaders in integrating sustainability into their supply chains due to more rigorous compliance requirements from their parent companies and international markets (Wu et al., 2013). In contrast, European companies are increasingly incorporating sustainability into their supply chains through compliance with EFRAG guidelines and the EU Taxonomy, which require detailed disclosures about the environmental impact of their operations, including those related to the supply chain (EFRAG, 2023b; Miandar et al., 2023). Figure 3.1 offers a summary of these frameworks and standards, particularly distinguishing between the GRI, ESRS, and International Sustainability Standards Board (ISSB), as well as the latter's frameworks and standards. The second part of Fig. 3.1 summarizes the impact of the frameworks on the materiality of climate-related disclosures.

Different legislations have taken different approaches to enforcing sustainability reporting, which express their diverging priorities and opportunities. European regulations tend to be more prescriptive and integrated, aiming for comprehensive disclosures that cover all aspects of sustainability. Chinese regulations are evolving to align more closely with global standards, while the U.S. continues to rely heavily on market-driven approaches. The following sections clarify the differences between regional legislative environments (Koberg & Longoni, 2019) in terms of how they attempt to implement sustainable practices amidst complex global supply chains.

3.3 The European Approach

The EU's commitment to sustainability and responsible business practices has led to a significant evolution in reporting standards, culminating in the adoption of the Corporate Sustainability Reporting Directive (CSRD) and the European Sustainability Reporting Standards (ESRS). This subsection explores the role played by the European Financial Reporting Advisory Group (EFRAG) in this

Comparison of Main Sustainability Disclosure Frameworks and Standards

				Consolidated into ISSB		
	GRI	ESRS	ISSB	Integrated Reporting Framework	SASB Standards	TCFD
Type of Guidance	Standards	Standards	Standards	Framework	Standards	Guidelines
Application	Voluntary	Mandatory for large companies and listed SMEs	Subject to national jurisdiction adoption	Voluntary	Voluntary	Voluntary
Coverage	Global	European Union (third countries in the future)	Global	Global	U.S., to be applicable globally	Global
Topics	Economic, environmental, and social activities and impacts	Environmental, social and governance	General Sustainability; Climate, Other topics to be added	Six capitals: financial, manufactured, intellectual, human, social, natural	Environment, social capital, human capital, business model & innovation, leadership & governance	Climate-related risks, opportunities, financial impacts, and scenario analysis
Sector Specific	No	Yes (forthcoming)	Yes	No	Yes	Yes
Target Audience	All stakeholders	All stakeholders	Investors	Providers of financial capital	Investors	Investors
Building Blocks		TCFD, GRI, CDP	TCFD, SASB, CDSB			
Materiality type	Impact materiality	Double-materiality (financial + impact materiality)	Single materiality (financial materiality)	Single materiality (financial materiality)	Single materiality (financial materiality)	Single materiality (financial materiality)
Materiality definition	Aspects that reflect the organization's significant economic, environmental, and social impacts; or that substantively influence the assessments and decisions of stakeholders	Impact on people or the environment and financial effects on undertaking over the short-, medium- and long-term time horizons.	Information is material if omitting, misstating or obscuring that information could reasonably be expected to influence decisions that the primary users of general-purpose financial reporting.	Matter that could substantively affect the organization's ability to create value in the short, medium, or long term.	A fact is material if there is a substantial likelihood that a reasonable investor would view its omission or misstatement as having significantly altered the total mix of information.	Public companies' legal obligation to disclose information in their financial filings—including material climate-related information.

Fig. 3.1 Comparison of the main sustainability disclosure frameworks and standards, and comparison matrix of climate-related disclosure requirements. *Source*: WB; IFC (2024), and The Evolution of Sustainability Disclosure: Comparing the 2024 SEC, ESRS, and ISSB Climate Disclosure

transformation, from the initial Non-Financial Reporting Directive (NFRD) to the latest developments under the CSRD and operationalization of the ESRS.

EFRAG has been instrumental in shaping the framework for sustainability reporting within the EU. As a key advisory body to the European Commission, EFRAG provides technical expertise and recommendations for developing

Comparison Matrix of Climate-related Disclosure Requirements

Comparison Matrix	SEC Climate Disclosure	ESRS	IFRS Sustainability Disclosure Standards (ISSB)
Jurisdiction	All SEC registrants, including non-US companies	All large companies in the EU subject to CSRD; listed companies on EU regulated markets except listed micro-enterprises.	ISSB Standards will be considered for adoption on a voluntary basis by individual jurisdictions
Likely date of the first report	Report covering FYB 2025 to report covering FYB 2027, depending on the type of SEC registrant	First reports expected by 2025 FY24 to FY28	Depending on the jurisdiction's implementation
Materiality	Financial	All stakeholders	Financial
Where to disclose	Annual report (some exceptions apply)	Management report	Annual report
Scope	Climate only	2 Cross-cutting core principles and 10 thematic ESG areas (5 on ENV, 4 on SOC and 1 GOV)	Currently – 1 standard on General sustainability and 1 on climate-related disclosure. Broader coverage is anticipated in the future.
GHG Reporting	Material Scope 1 and/or Scope 2, if the company is not exempted	Scope 1 &2. Scope 3 with a phased-in period	Scope 1, 2 and3. Scope 3 with possible 1 year relief period and guidance
Assurance Requirements	Attestation of GHG emissions required, if the registrant is required to report Scope 1 and/ or Scope 2 emissions	Limited assurance requirements are expected within three years after implementation and reasonable assurance after six years	To be determined by the implementing jurisdiction

Adapted from: WB and IFC, 2024, and The Evolution of Sustainability Disclosure: Comparing the 2024 SEC, ESRS, and ISSB

Note: **CSRD** = Corporate Sustainability Reporting Directive; **EFRAG** = European Financial Reporting Advisory Group; **ESG** = environmental, social, and governance; **ESRS** = European Sustainability Reporting Standards; **EU** = European Union; **FY** = Fiscal Year; **ISSB** = International Sustainability Standards Board; **SEC** = Security and Exchange Commission.

Source: WB and IFC, 2024, and The Evolution of Sustainability Disclosure: Comparing the 2024 SEC, ESRS, and ISSB Climate Disclosures

Fig. 3.1 (continued)

reporting standards that are not only comprehensive but also practicable across various sectors and enterprise sizes.

Initially established in 2014, the NFRD provided the foundation for non-financial reporting in the EU, with a focus on social and environmental impacts. However, to address the evolving needs of stakeholders and the growing demand for more detailed and reliable data, policymakers replaced the NFRD with a more stringent CSRD in 2023. This directive enhances the scope and depth of reporting in order to create greater transparency and comparability across companies and industries.

With the CSRD in place, the European Commission adopted the ESRS in July 2023. These standards are designed to make sustainability reporting consistent across the EU, allowing investors, consumers, and other stakeholders to more easily evaluate companies' sustainability performances. The ESRS covers a broad spectrum of ESG factors and is integral to the EU's broader Sustainable Finance Framework, which includes the EU Taxonomy for Sustainable Activities.

The following sections of the chapter analyze these topics, examining how EFRAG's regulatory framework, shaped by the adoption of CSRD and ESRS, is influencing not only European businesses but also setting a benchmark for sustainability reporting at the international level that expands on existing frameworks (e.g., GRI; Fig. 3.2).

3.3.1 *The Non-Financial Reporting Directive (NFRD)*

The European Union's Non-Financial Reporting Directive (NFRD) 2014/95/EU was instituted to enhance transparency in large companies' reporting of environmental, social, and governance issues (European Commission, 2014, p. 1). This legislation has had significant implications for supply chain management practices across various industries. This section explores the effects of NFRD on supply chain practices, utilizing a selection of academic studies to identify key topics and discussing the limitations of this regulatory approach.

According to Biondi et al. (2020), the NFRD has encouraged companies to adopt more comprehensive reporting frameworks, like the International Integrated Reporting Framework. This transition has led to increased transparency but risks becoming another reporting façade if it is not accompanied by a genuine commitment to sustainability. After all, surface-level compliance can obscure the deeper, systemic changes needed to evolve supply chain practices. Furthermore, this issue is compounded by the fact that EU member states differ in how they translate the regulations to their own context (Aureli et al., 2019). In line with this reasoning, La Torre et al. (2018) discussed the challenges of harmonizing non-financial reporting across Europe: While the directive aims to standardize practices, significant variations still exist due to different interpretations and implementations at the national

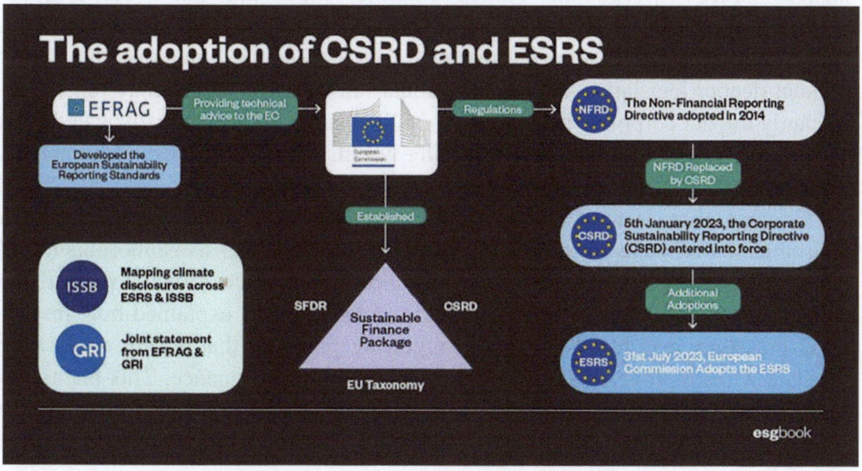

Fig. 3.2 The European approach to sustainable finance and sustainability reporting. *Source*: ESG Book's Reporting Exchange (2023)

level (p. 612). This lack of uniformity can lead to inconsistencies in how supply chain impacts are reported and managed across borders. Even though several issues are linked with the national transposition process and applicational level (Chelli et al., 2018; Moggi et al., 2023), there are others—such as the quality of reporting—that remain a challenge at the organizational level and thus hinder the development of uniform performance measures (Pagani et al., 2021). Accordingly, Traxler et al. (2023) assessed whether NFRD improved the quality of reporting and found mixed results. Some companies have indeed enhanced their disclosure practices; others, however, have merely complied with minimum standards without truly integrating sustainability into their operational or strategic frameworks. This has led to discussions about the motivations for disclosure, namely, between voluntary and mandatory reporting (Aureli et al., 2020; Baboukardos et al., 2023).

Hoffmann et al. (2018) highlighted the tension between the mandatory and voluntary aspects of non-financial reporting in Germany. While mandatory reporting has raised the floor of sustainability reporting, it has not necessarily shifted the ceiling; that is, best practices remain largely driven by market leaders rather than legislative requirements (p. 59). Scholars have found that market leaders often trigger a change in business model orientation, which impacts not only firms' internal context but also their supply chains, given their role as focal companies (see the recalled framework of Seuring and Müller, 2008). The impact of the market leader is therefore twofold: On the one hand, it drives the standardization of sustainability practices across the supply chain; on the other hand, it encourages the improvement of internal processes. Internal process quality improvements can likewise be achieved through an external audit of the reporting process. This, in turn, leads to enhanced quality and reliability of the disclosed data. The reliability of the information provided has also been associated with higher market value, thanks to the assurance process of these reports (Fazzini & Dal Maso, 2016; Junior et al., 2014). That said, changes induced by regulatory bodies or market leaders may not be sufficient for triggering a substantial rethinking or adaptation of current systems. In this vein, Monciardini et al. (2020) argued for structural changes in the regulatory framework: Without deeper legislative and organizational reforms, the NFRD may fail to significantly impact corporate behaviors in supply chains. The authors advocated for a blueprint that not only mandates reporting but also integrates these requirements into a broader corporate governance framework.

Of course, there is a risk that external demands for adaptation—especially in an environment that lacks sustainability orientation—will lead to isomorphism and superficial compliance strategies. On this topic, Stefanescu (2022) explored the concept of isomorphism within the context of NFRD. She explained how institutional pressures lead to similarities in reporting practices that do not necessarily correlate with improvements in actual sustainability performance. This phenomenon can result in companies adopting similar sustainability practices for legitimacy rather than operational effectiveness (see Sect. 2.1 for more discussion on this aspect).

Scholars have uncovered mixed empirical evidence that firms are struggling with substantive adaptation. For example, Raucci and Tarquinio (2020) looked at the

Italian context and found that sustainability performance indicators are being increasingly integrated into reports due to the NFRD; however, there is still a gap in how these indicators translate into concrete sustainability improvements in supply chain actions, such as the "negative social impacts in the supply chain" (p. 12). Similarly, Muserra et al. (2020) highlighted the discrepancy between policy intentions and actual outcomes in Italy, noting that some firms view the directive as a compliance exercise rather than an opportunity to enhance sustainability. They thus take advantage of the flexibility in the standard choice to resist true standardization of measures and comparable results. On the other hand, Muserra et al.'s (2020) country-level analysis of Italy found that NFRD "is considered an effective way to promote non-financial disclosure without imposing further compliance obligations on companies" (p. 30).

The EU's NFRD has pushed European companies toward greater transparency in their supply chain practices. However, the academic literature suggests that without a stronger emphasis on the quality of reporting and genuine integration of sustainability into business operations, the directive may not fulfil its potential to drive profound environmental and social change. Future regulatory efforts must address these limitations by fostering not just compliance but a deep-rooted commitment to sustainability throughout corporate supply chains. The current Corporate Social Responsibility Directive could be a further step in that direction.

3.3.2 The Corporate Social Responsibility Directive (CSRD)

The NFRD represents the first attempt to integrate various national regulations, but the varied applications and integrations of the regulation have led companies to output information that is difficult to compare. Relatedly, this fragmentation (and occasional arbitrariness) cannot fully satisfy the informational needs of diverse stakeholders. For these reasons, the European Commission (2021) noted that "the information that companies report is not sufficient. Reports often omit the information that investors and other stakeholders think is important. Reported information can be hard to compare from company to company, and users of the information are often unsure whether they can trust it." This issue not only affects companies' level of perceived legitimacy but also their competitiveness compared to other less regulated contexts that have access to the European single market (Odobaša & Marošević, 2023). The underlying idea, shared by the literature, is that a higher level of information disclosure can contribute to risk management (Veltri, 2020) and, by extension, greater investor confidence (Lykkesfeldt & Kjaergaard, 2022).

After several drafts and public consultations, the ESRS went into force in the reporting year 2024 for a progressive number of firms (EFRAG, 2022a). As noted by Čufar et al. (2024), this process is important for transitioning more companies in Europe from voluntary to binding non-financial reporting. The initiatives outlined in the text—such as the Sustainable Finance Platform and various voluntary mapping and reporting frameworks—are directly connected to the CSRD. The CSRD aims to enhance and standardize sustainability reporting across the EU, building on the foundation laid by the Sustainable Finance Platform (European Union Platform on Sustainable Finance, 2021). By mandating more comprehensive and comparable disclosures of sustainability impacts, risks, and opportunities, the CSRD seeks to make the financial sector a key driver of firms' sustainability transformation. This alignment ensures that reporting frameworks—such as the Taskforce on Climate-Related Financial Disclosures, the Global Reporting Initiative (GRI), and the OECD Guidelines for Multinational Enterprises—are utilized within a consistent regulatory framework. This will ideally serve to improve transparency and foster greater investor confidence in sustainability efforts (Andersson & Arvidsson, 2023).

Despite the deadline of July 6, 2024, numerous EU member states have yet to incorporate the CSRD into their national legislation. According to the Ropes and Gray (2024) EU CSRD Transposition Tracker, only 11 countries have successfully implemented (at least partially) the required legislation by July 2024. These countries were Denmark, Finland, France, Hungary, Ireland, Lithuania, Romania, Slovakia, Sweden, Liechtenstein, and Norway. Eight other countries proposed legislation, while four conducted consultations (Fig. 3.3). In fact, seven countries (Austria, Belgium, Germany, Greece, Malta, Portugal, and Iceland) had neither introduced legislation nor initiated consultations by the time of the application of the directive. Notably, on July 24, the German Federal Cabinet approved the government draft for transposing the CSRD into German law, which was subsequently published on the website of the Federal Ministry of Justice.[1]

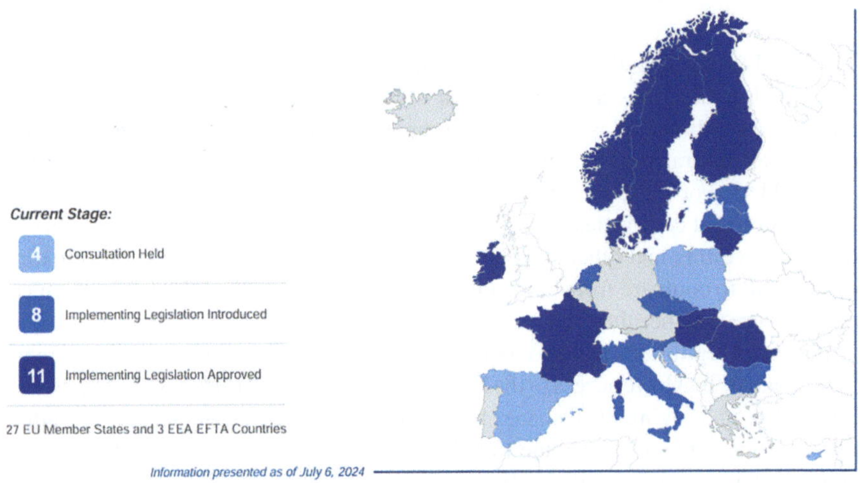

Fig. 3.3 Transposition of the CSRD into national law by the July 6, 2024 deadline. *Source*: Adapted from Ropes and Gray CSRD Transposition Tracker (July 6, 2024)

[1]Further information can be found at: https://www.bmj.de/SharedDocs/Gesetzgebungsverfahren/DE/2024_CSRD_UmsG.html?nn=110490.

Figure 3.4 presents the most updated version of the tracker as of October 31, 2024. While Spain continues to hold consultations, most of the EU members have at least introduced, if not passed, legislation.

3.3.3 *The European Sustainability Reporting Standards (ESRS)*

The European Financial Reporting Advisory Group (EFRAG) and the European Sustainability Reporting Standards (ESRS) are set to define the landscape of European sustainability reporting for the near future. EFRAG was established to provide expert advice to the European Commission on accounting matters but has expanded its role to include the development of the ESRS under the new CSRD. EFRAG's role involves not only advising on corporate reporting but also developing standards that address wide-ranging ESG issues. Its recent initiatives—like the EFRAG IG 3 List of ESRS Data Points—illustrate an ambitious step toward detailed and standardized reporting (EFRAG, 2022b). This initiative highlights the specific data points that companies are expected to disclose to ensure thorough coverage of pertinent sustainability information. The ESRS, guided by the frameworks developed by EFRAG, seeks to standardize sustainability reporting across EU member states so that companies can provide comparable information regarding their environmental and social impacts (Parrondo, 2024). While retracing the entire process of development and approval for the ESRS is beyond the scope of this chapter, Fig. 3.5 positions the ESRS within the broader European intervention framework. It also outlines all the constituent elements of the topics covered by the standard.

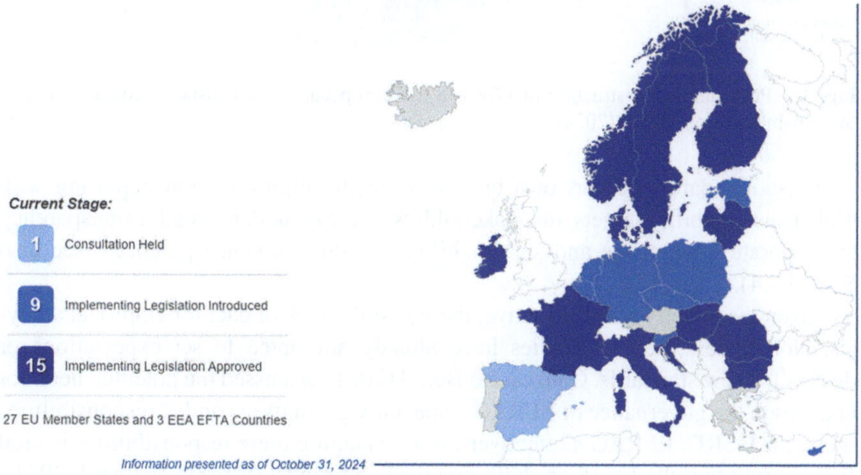

Fig. 3.4 Transposition of the CSRD into national law on October 31, 2024. *Source*: Adapted from Ropes and Gray CSRD Transposition Tracker (October 31, 2024)

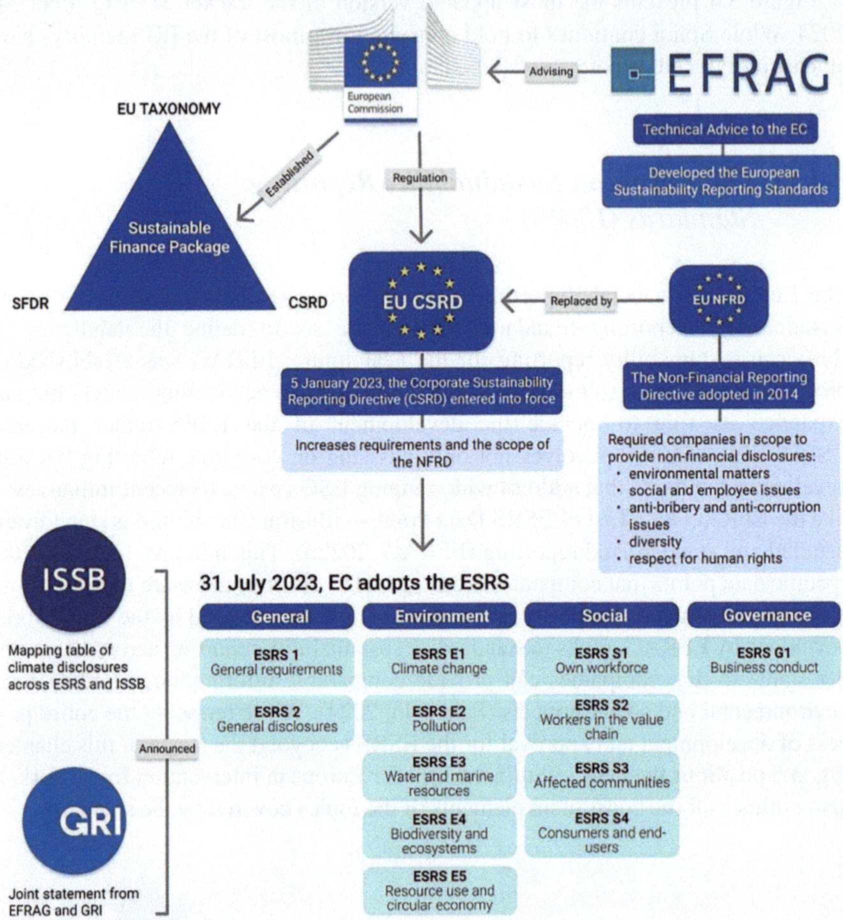

Fig. 3.5 Positioning and structure of ESRS in the European Union legislative context. *Source*: Sustainability Infographics (2024)

These standards are part of a broader effort to align corporate reporting with global sustainability targets and stakeholders' increasing demands for transparency in corporate governance and sustainability and due diligence practices (see also Sect. 3.3.4).

Given the novelty of the directive, there is still a lack of data for empirical analysis. Nonetheless, several studies have already attempted to set expectations or desires for new standards. Gauzès and Board (2021) discussed the potential need for changes in the governance of EFRAG, especially given its expanded responsibilities under the CSRD. EFRAG's effectiveness in managing these responsibilities is vital for ensuring that the ESRS are both practical and robust (Gauzès & Board, 2021). Dai et al. (2021b) explored the interplay between institutional pressures (e.g., those potentially exerted by ESRS in the near future) and the development of sustainability capabilities. They argue that such regulatory frameworks significantly encourage organizations to enhance their sustainability practices, particularly in complex supply chains (Dai et al., 2021a). This is crucial, as supply chains often span multiple regulatory environments and hence need cohesive policies to manage sustainability effectively. Kazancoglu et al. (2021) highlighted how such policies can encourage circular economy practices within supply chains, which entail promoting sustainability from production to end-of-life management. In their view, "governments' role should not only be limited to legislation and regulations but also act as the motivator and supporter by establishing standards, infrastructural investments, and incentives for the transition to CE for enhanced corporate environmental management. Also, legal arrangements must be clear, formal, coherent, and compulsory to achieve higher corporate environmental performance levels" (Kazancoglu et al., 2021, p. 604).

While the EFRG hopes to promote greater organizational uniformity within the European context, other regions do not currently offer the same comparability. In their globe-spanning research, Tsai et al. (2021) found that European companies are leading the disclosure of sustainable supply chain management information in Asia and Oceania, followed by North Americans. Europe is at the forefront of implementing advanced sustainability practices, although the degree of implementation varies significantly. Antonini and Gomez-Conde (2024) discussed the unintended consequences of the EU non-financial reporting directive, noting that while the directive aims to enhance transparency, it may also lead to compliance-focused behavior that does not necessarily translate into real sustainability gains.

Evidently, there is a delicate balance between enforcing compliance and fostering 'substantial' sustainability integration, particularly in the domains of human rights, environmental management, and sustainability innovation. Martin-Ortega (2022) discussed the shift from corporate-led disclosure to a rights-based approach in global supply chains, emphasizing the importance of transparency in human rights practices. This perspective is particularly relevant to the ESRS, as the standards aim to improve transparency not just in environmental issues but also in social governance aspects (e.g., human rights). The ESRS promotes a framework in which rights-related disclosures are mandatory, helping stakeholders gain a clearer understanding of how corporate practices impact communities and workers (Martin-Ortega, 2022). This aligns with the ESRS's broader goal of ensuring that companies provide comprehensive reports on their sustainability practices, including how they manage their supply chains ethically and responsibly. The expected effect of the ESRS is therefore twofold: On the one hand, it encourages companies to engage with the issue of fostering attention toward the ethics of relationships within supply chains; on the other, it sets out (minimum) levels that must be met to satisfy the standard. To this end, companies need to integrate ESRS with their environmental management system as a relevant antecedent in order to obtain a proper implementation of the standards. In line with this reasoning, the CDP (2022) Global Supply Chain Report 2021 and research by Daddi et al. (2021) both highlight the importance of integrating environmental management systems into supply chain practices, thus expanding the reporting boundaries from the micro-level to the meso-level.

The CDP report underscored the increasing demands from consumers and regulators for detailed disclosures on environmental impacts and the strategies that companies employ to mitigate these impacts (CDP, 2022). Similarly, Daddi et al. (2021) explored how green supply chain management capabilities enhance the internalization of environmental management systems, leading to improved organizational performance. These insights suggest that the ESRS, by standardizing such reporting requirements, can significantly influence companies to adopt more sustainable (and likely more resilient) supply chain practices. Using simulation studies, Ivanov (2018) observed that resilient supply chains can better adapt to disruptions while maintaining their commitment to sustainability goals. Thus, ESRS encourages firms to not only report on their immediate sustainability practices but also demonstrate how these innovative practices contribute to the long-term resilience of their supply chains.

To better understand firms' readiness for these changes, EFRAG conducted a study on 28 companies with a focus on the implementation of the ESRS. The study, which relied on interviews and surveys, exposed several important trends and challenges faced by these companies. This subsection reports the evidence concerning value chain analysis (See Sect. 1.5 for further details) and the double materiality assessment approach (See Sect. 2.4). Unsurprisingly, value chain analysis has been identified as the least mature area for ESRS implementation. According to the survey (EFRAG, 2024, p. 6):

- "~90% of the undertakings are still working to refine their value chain(s) mapping, looking for the right balance of granularity.
- ~45% of the undertakings have already adopted a more granular mapping of their value chain (i.e., going beyond the high-level upstream, downstream, and own operations only).
- While some undertakings prioritized direct business relationships (Tier 1) for PATs11 and IROs12 for the initial implementation of ESRS, several are already going beyond Tier 1, especially non-financial institutions."

The concept of Double Materiality Assessment—integrating both financial and impact materiality—is another key topic in ESG reporting. According to a previous study (EFRAG, 2024, p. 6):

- "~85% of the undertakings intend to integrate ESG reporting and the result of the Double Materiality Assessment into business strategy and decision-making.
- ~70% of the undertakings have already started to deploy an objective evidence-based approach to Double Materiality Assessment, using a combination of data sources, enhanced by the inputs of internal experts and stakeholders."

A significant proportion of companies are adopting structured, data-driven approaches to double materiality, but there is a noticeable gap in the actual implementation of these practices. The use of evidence-based methods is promising, but the practical challenges of data integration and comprehensive value chain analysis need to be addressed to ensure full compliance and effective reporting. This anecdotal evidence reflects the legislator's attempt to link its regulatory role to its effectiveness in informing managerial practices at the firm level.

The July revisions pursue simplification without abandoning the core logic of the CSRD. Specifically, EFRAG concentrates on cross-cutting mechanics in ESRS 2, streamlines drafting conventions, and relocates many illustrations into Non-Mandatory Illustrative Guidance (NMIG), while eliminating "may-disclose" items from the body of the standards to limit duplication and boilerplate (EFRAG, 2025).

Consultation is open through September 29, 2025, underscoring that this is an exposure text and inviting evidence-based feedback on the burden–usefulness balance (EFRAG, 2025). In parallel, public commentary frames drafts as the maximum simplification consistent with EU objectives, cautioning against further cuts that could weaken decision-useful reporting (Frank Bold, 2025b). Within ESRS 2, the re-architecture clarifies narrative flow by separating the description of material impacts, risks, and opportunities (IRO-2) from the interaction with strategy and financial effects (SBM-3), with application requirements that explicitly warn against generic, standardized disclosures and direct preparers to company-specific processes and inputs (EFRAG, 2023a, 2025).

The same application guidance links site and asset screening to a place-based context and recognizes the TNFD LEAP sequence as a useful reference, reinforcing an IRO-first approach that then anchors policies, actions, targets, and metrics (EFRAG, 2025). Independent analysis broadly welcomes this structural shift—while noting that removing topic-level details from topical standards reduces immediate on-page cues for practice and may invite superficial narrative if firms do not truly start from IROs (Frank Bold, 2025a).

A central controversy concerns some anticipated financial effects. EFRAG's draft introduces two pathways: Option 1, which permits qualitative explanations where quantification is not reasonably determinable, and Option 2, which would allow purely qualitative disclosure. As written, the application guidance to Option 1 still expects "reasonable and supportable" information and ties anticipated effects on planning horizons and capital allocation, thereby maintaining a bridge to financial materiality expectations (EFRAG, 2025). Civil society reviewers argue that relief should not extend beyond Option 1, warning that removing quantification could obscure exposures precisely where investors seek forward-looking discipline (Frank Bold, 2025a; Frank Bold, 2025b). In practice, firms that cannot quantify in year one should disclose the basis for this judgment and document a time-bound plan to build capability—aligning with the spirit of the new ESRS 2 architecture (EFRAG, 2025).

The exposure drafts also provide new relief mechanisms (e.g., the omission of certain metrics where obtaining data would entail "undue cost or effort") and permit the exclusion of data from joint operations without operational control. While these provisions can reduce near-term burden, critics note the absence of mandatory improvement plans or sunset clauses, with potential consequences for comparability if such reliefs are overused (Frank Bold, 2025a). Here again, the governance remedy is procedural: contemporaneous documentation of the rationale for relief, clear sequencing toward fuller coverage, and explicit cross-references to the relevant policies, actions, and targets under the ESRS 2 general disclosures (EFRAG, 2025).

Across topical standards, environmental changes illustrate the trade-off between readability and granularity. E1 (climate) keeps the transition plan backbone but trims embedded guidance; E1–8 introduces flexibility by allowing an operational-control basis for Scope 1–2 where needed for fair presentation, alongside financial control—flexibility that can be useful but also risks under-reporting if applied incautiously (Frank Bold, 2025a). For E2–E4, the pruning of detailed PAT elements and the removal of site-screening cues and the marine-resources sub-topic make texts leaner but shift the burden of decision-useful specificity to NMIG and external frameworks (e.g., TNFD, SBTN) (Frank Bold, 2025a). The public commentary converges with this diagnosis, acknowledging the improvement in structure, while warning that losing measurement exemplars—particularly for biodiversity—creates practice gaps that need to be filled through guidance (Frank Bold, 2025b).

Social pillars mirror this pattern. The drafts refocus S-disclosures on actions, effectiveness, and engagement with affected stakeholders in a manner closer to due diligence logic, yet adjustments to metrics carry policy salience. The proposed methodology for S1–9 (adequate wages) leans on ILO wage-setting principles and may be impractical without access to credible living-wage benchmarks, while the removal of the family related leave usage metric in S1–14 reduces visibility into work–life balance and gender-equity dynamics that many users monitor (Frank Bold, 2025a).

Finally, governance is tighter and better aligned with cross-cutting conventions, but one deletion matters for supply chain fairness: dropping invoice payment times from G1 diminishes routine transparency on buyer–supplier practices, since the retained litigation metric rarely captures day-to-day stress across SME relationships (Frank Bold, 2025a). This is precisely where an IRO-first reading—supported by ESRS 2's application guidance against boilerplates—can sustain decision usefulness despite a leaner set of data points (EFRAG, 2025; Frank Bold, 2025a).

In sum, the July drafts strengthen internal coherence and reduce duplication while transferring part of the granularity to guidance and external frameworks. The net effect will depend on implementation choices: Firms that genuinely anchor disclosures in entity-specific IROs, explain and time-box any use of reliefs, and maintain progressive quantification of anticipated effects will likely preserve comparability and insight, and those that do not risk reverting to generic narrative (EFRAG, 2025; Frank Bold, 2025a; Frank Bold, 2025b). From this perspective, supply chain management not only aligns with the disclosure requirements of the ESRS but also enables effective due diligence processes, ensuring that companies can maintain responsible and sustainable operations (Le et al., 2022).

3.3.4 *The Corporate Sustainability Due Diligence Directive (CSDDD)*

The Corporate Sustainability Due Diligence Directive (CSDDD) represents a significant regulatory shift toward mandating corporate compliance in addressing adverse human rights and environmental impacts (Directive (EU) 2024/1760 of the European Parliament and of the Council). The directive enforces stringent obligations on companies (European Commission, 2024, p. 4), such as requiring large companies to integrate due diligence into their policies and risk management systems (European Commission, 2024, p. 5). While the EFRAG and ESRS frameworks aim to standardize and improve sustainability reporting across Europe, their implementation produces significant externalities in how companies manage their supply chains. These regulations do not merely define what should be reported; they also reshape the supply chain itself by promoting processes aligned with the principles of the circular economy, both within and beyond the boundaries of the EU. The new CSDDD is closely aligned with the path proposed by the United Nations (UN), as emphasized by Higham (2019) and Rasche and Waddock (2021). The latter authors highlighted how the UN Guiding Principles on Business and Human Rights (UNGPs) provide a foundational framework for CSR research, emphasizing corporate responsibility in respecting human rights. More recently, Rogerson et al. (2024) provided evidence that the largest 100 EU-listed firms are neglecting the topic of human rights in their due diligence disclosure and (more broadly) managerial practices. Higham (2019) discussed the standardization role of the UNGPs, underscoring their importance in risk governance and corporate accountability. Moreover, he has also illustrated the critical infrastructure that UNGPs provide for embedding robust CSR practices within corporate operations.

According to Mak (2022), several regulations have been developed at the national level to engage firms in corporate sustainability and due diligence. These regulations go beyond mere formalities and aim to embed substantial CSR practices into corporate operations. Mak (2022) emphasized that these national measures often serve as more stringent and context-specific extensions of the UN Guiding Principles, seeking to ensure that firms are not just 'ticking boxes' but genuinely committing to sustainable business practices in their 'chain of activities' (European Commission, 2024, p. 2). In addition, the CSDDD aligns closely with the OECD Guidelines for Multinational Enterprises and the ILO Tripartite Declaration of Principles concerning Multinational Enterprises and Social Policy. The OECD Guidelines, updated in 2023, require enterprises to conduct due diligence to address human rights and environmental impacts (OECD, 2023). This alignment ensures that the CSDDD's requirements are grounded in internationally recognized standards, promoting global coherence in corporate sustainability and especially corporate governance (Torelli et al., 2025). Similarly, the ILO Declaration, amended in 2022, embeds the concept of human rights due diligence in the broader urgency for companies to respect and address labor and human rights throughout their operations (ILO, 2022). By integrating these principles, the CSDDD reinforces global standards for human rights protection. The directive mandates comprehensive due diligence processes, ensuring that companies identify, prevent, mitigate, and account for adverse impacts on their operations and supply chains. This harmonized approach not only strengthens ethical business practices within the EU but also facilitates smoother transitions for multinational enterprises familiar with the OECD and ILO frameworks.

The CSDDD initiative supplements national legislation like France's Corporate Duty of Vigilance Law,[2] Italy's Due Diligence Laws,[3] and the Netherlands' Child Labor Due Diligence Law.[4] The introduction of the proposed Directive also coincides with the advancement of due diligence laws in countries such as Germany[5] and the Netherlands.[6] This engagement is crucial for addressing specific local challenges (i.e., child labor or forced labor, mainly in extra-EU contexts) and fostering a culture of accountability and transparency in corporate conduct.

Since companies are obliged to adapt to these regulations, they are compelled to reevaluate and even redesign their supply chain operations to meet new standards of sustainability reporting and performance while facing the complexity embedded in this operation (Villiers, 2022). This regulatory push toward greater transparency and accountability in sustainability practices is reshaping corporate behavior, driving a shift toward more sustainable business practices across Europe, and potentially limiting the risk of labor rights violations across layers of the supply chain (Hurt et al., 2023).

The European Coalition for Corporate Justice (2024) proposed an effective visualization of the directive's structure (Fig. 3.6), which reports its expected and estimated impacts on human rights and environmental protection obligations.

This integration pushes companies to systematically identify, mitigate, and end adverse impacts within their operations and supply chains. The directive's mandatory nature closes the gaps left by voluntary frameworks, providing a harmonized legal standard across the EU (European Commission, 2024, p. 5).

The obligations under the CSDDD are comprehensive. Companies must identify adverse human rights and environmental impacts in their operations and those of their subsidiaries and business partners. This includes addressing the impacts based on their severity and likelihood. Companies are required to implement prevention

---

[2] Law nr. 2017–399, 27 March 2017. More information can be retrieved at: https://www.legifrance.gouv.fr/jorf/id/JORFTEXT000034290626/.

[3] Legislative Decree Nr. 231/2001, Gazzetta Ufficiale Nr. 140, 19 June 2001, https://web.camera.it/parlam/leggi/deleghe/01231dl.htm, Amended by Legislative Decrees Nr. 184/2021, https://www.gazzettaufficiale.it/eli/id/2021/11/29/21G00200/sg, and No. 195/2021, https://www.gazzettaufficiale.it/eli/id/2021/11/30/21G00203/sg.

[4] Netherlands' Child Labour Due Diligence Law, 14 May 2019, Nr. 401. More information can be retrieved at: https://www.eerstekamer.nl/wetsvoorstel/34506_initiatiefvoorstel_kuiken.

[5] German Supply Chain Due Diligence Act, adopted in June 2021 and entered into force on 1 January 2023, Bundesgesetzblatt 2021. More information can be retrieved at: https://www.bundesanzeiger.de/pub/publication/0zcDT5oLsZVqBzPDiTi.

[6] Netherlands' Child Labour Due Diligence Law, approved on 14 May 2019, Nr. 401, https://zoek.officielebekendmakingen.nl/stb-2019-401.html.

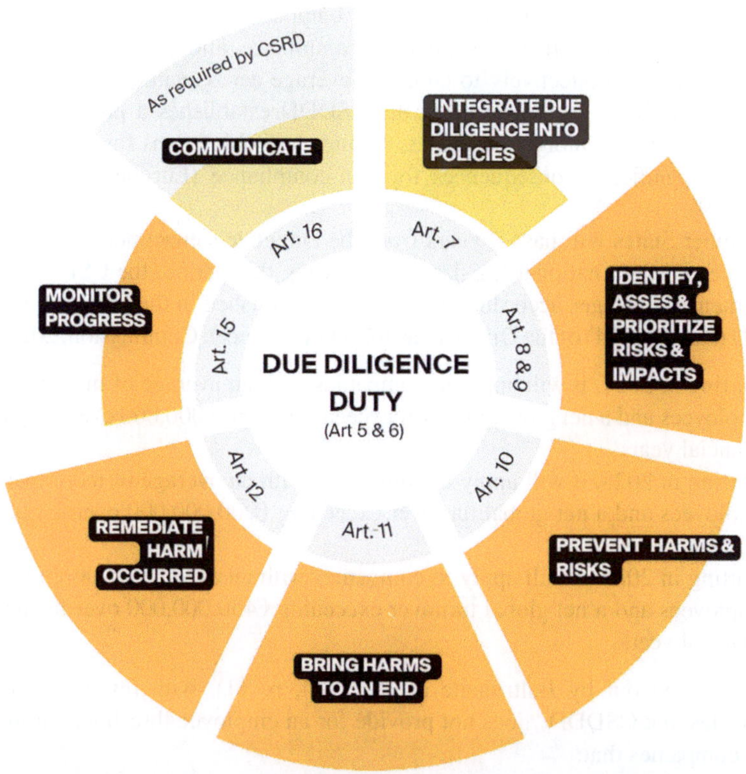

Fig. 3.6 Impact of the CSDDD on human rights and environmental protection obligations imposed on companies. *Source*: European Coalition for Corporate Justice (2024)

and corrective action plans, obtain contractual assurances from business partners, and make necessary financial and non-financial investments. Additionally, companies must engage with stakeholders, establish complaints and notification procedures, monitor the effectiveness of due diligence measures, and publicly communicate their efforts as per the CSRD and ESRS previously discussed.

A critical aspect of the directive is the enforcement regime, which includes both administrative and civil liabilities. Competent national authorities enforce compliance through injunctions and sanctions, ensuring that penalties are effective, dissuasive, and proportionate. Companies failing to comply may face fines and other corrective measures. Civil liability provisions mean that companies can be held accountable for damages caused by their failure to comply with due diligence obligations, necessitating compensation for victims.

The directive also addresses potential externalities, particularly in developing countries. The directive's anticipated benefits include improved human rights protection, more sustainable business operations, increased adherence to international standards, and greater investments in EU companies. These positive externalities will ideally generate better living conditions and social justice for individuals in these regions. Furthermore, the directive emphasizes the role of industry and multi-stakeholder initiatives in supporting compliance, allowing companies to pool resources and act collectively to enhance leverage across value chains (European Commission, 2024, p. 14). Overall, the CSDDD establishes a potentially robust framework for mandatory compliance, setting clear obligations for companies and outlining significant consequences for non-compliance (European Commission, 2024, p. 9).

Member States will have 2 years from the Directive's enactment to incorporate its framework into national legislation. Following this period, the CSDDD will be implemented in stages according to the version published in the Official Journal of the EU (July 5, 2024) of the Directive as follows (European Commission, 2024, p. 5):

- Starting in 2027, it will apply to companies with an average of more than 5000 employees and a net global turnover exceeding €1,500,000,000 over the past two financial years.
- Starting in 2028, it will apply to companies with an average of more than 3000 employees and a net global turnover exceeding €900,000,000 over the past two financial years.
- Starting in 2029, it will apply to companies with an average of more than 1000 employees and a net global turnover exceeding €450,000,000 over the past two financial years.

As pointed out by Baltrunaite et al. (2024, p. 11), with respect to non-EU companies, the CSDDD "does not provide for an employee threshold but includes those companies that:

- had a net turnover in the EU of more than €450,000,000 in the last two financial years; or,
- did not reach the threshold as referred to in point (a) but are the ultimate parent companies of groups that on a consolidated basis reached that threshold in the last two financial years."

Thus, the CSDDD is set to impact not only the approximately 6000 companies within the European context but also a large number of foreign companies with significant turnover exposure within the EU. The next subsection discusses the impact of European legislation on these foreign companies' operations.

3.4 Externalities of European Regulations

The European NFRD has not only influenced the direct reporting practices of companies but also exerted broader externalities on supply chain management. Given the globalization of current supply chains, the CSRD and CSDDD will introduce potential spillover effects (Gold & Wieland, 2024) on many companies, including those outside the European scope (see the right side in Fig. 5.1). This section outlines several key studies on prior interventions (Balon, 2020; Dai et al.,

2021b; Kazancoglu et al., 2021; Tsai et al., 2021). In fact, these perspectives offer preliminary insight into the potential outcomes of current legislation. Specifically, it anticipates how regulatory efforts may influence not only compliance behavior but also foster a deeper integration of sustainability principles into operational strategies (Yang et al., 2024).

Balon (2020) highlighted how NFRD has pushed companies to adopt green supply chain management practices under growing regulatory and stakeholder pressures. This shift is emblematic of broader market trends, where companies are increasingly held accountable for their environmental footprints (Mızrak & Cevher, 2023; Sarkis et al., 2016). Due to external regulations, there will be a decrease in flexibility and adaptability (directly) at the firm level (Antonini & Gomez-Conde, 2024), as well as (indirectly) at the supply chain level. Dai et al. (2021b) explored the dynamics between these institutional pressures, like those exerted by the NFRD, and the development of sustainability capabilities within organizations. Their findings indicate that such regulatory pressures significantly influence companies to develop capabilities that support sustainable supply chain practices, thereby creating a more resilient and responsive supply chain network (Dai et al., 2021b). Therefore, these regulations not only define what needs to be reported but also affect the very configuration of the supply chain itself. In this vein, Kazancoglu et al. (2021) designed a framework for improving corporate environmental management in supply chains in line with circular economy principles. They argue that policies such as the NFRD catalyze the adoption of circular economy practices by embedding these requirements within the regulatory framework (Kazancoglu et al., 2021). However, the impact of formalization is not uniformly positive: By focusing on and highlighting certain aspects of reporting, the legislation leaves room for firms to engage in opportunistic behaviors or merely comply with formal obligations. Antonini and Gomez-Conde (2024) discussed the unintended consequences of NFRD, particularly how its emphasis on compliance can sometimes lead to box-ticking behaviors rather than genuine sustainability improvements. Without proper implementation and enforcement, they argue that CSRD and CSDDD might lead to superficial compliance that does not translate into real environmental benefits or innovation propensity (Antonini et al., 2024). The EU has (partially) attempted to mitigate this issue through a 'comply or explain' approach (Cuomo et al., 2024).

Another issue to consider regarding externalities is the magnitude of their influence. Tsai et al. (2021) conducted an analysis of sustainable supply chain management trends across different world regions, seeking to discern how global directives like the NFRD influence these trends. Their research showed that such regulations have a significant impact on raising the baseline for sustainability practices across regions, although the specific impacts can vary widely based on local implementation and cultural factors (Tsai et al., 2021). Other scholars—such as Maas and Sampers (2020) and Monciardini et al. (2020)—have called for a rethinking of the regulatory framework: not merely to mandate reporting, but to drive substantive changes in corporate sustainability practices. In particular, Monciardini et al. (2020) argued for structural regulatory changes that align more closely with the complexities of modern supply chains and firms' varying capacities to respond to these directives (Monciardini et al., 2020, p. 11).

Guo et al. (2025) documented that green-innovation practices travel "upstream": When a focal company invests heavily in green innovation (and especially when its technological capabilities outpace those of its suppliers), those suppliers respond by intensifying their own green-innovation activities. The authors traced this diffusion across a large Chinese buyer–supplier panel (2009–2023) and showed that the spillover is strongest in non-state-owned and lightly regulated industries, indicating that market-based incentives (rather than coercive regulation) lie at the heart of the effect. Guo et al. (2025) further observed the same international and supply chain forces that push suppliers to increase focal firms' likelihood of disclosing or patenting green inventions, suggesting a shared mechanism along the chain. Taken together, the evidence supports the view that multinational customers transmit demands for credible environmental information and innovation through relational ties—a finding consistent with broader work on information-based imitation and knowledge transfer in supply networks (Dai et al., 2021a, 2021b; Isaksson et al., 2016; Lieberman & Asaba, 2006; Tang et al., 2023).

To date, scholars have debated how the European NFRD has shaped the integration of sustainability issues into supply chain practices across Europe. Moreover, from the current European policy perspective, these results imply that regulations such as the CSRD and the forthcoming CSDDD could amplify positive externalities by formalizing data-sharing templates and due diligence routines. Mandatory value chain reporting is likely to strengthen the information-exchange channel highlighted by Guo et al. (2025), prompting lead firms to equip smaller suppliers with measurement tools and technical guidance—similar to what Song et al. (2024) reported for carbon emissions transparency. At the same time, policymakers must guard against unintended side effects (e.g., supplier switching or "cut-and-run" behavior) by ensuring proportional requirements for SMEs and promoting collaborative remediation rather than pure compliance pressure.

In that regard, Villena and Gioia (2020) set out an idealized view of sustainable supply chain management, rooted in the practices of multinational enterprises. In this view, a focal firm commits to working only with suppliers who meet defined social and environmental expectations. It not only holds its first-tier suppliers to those standards but also expects them to pass the same requirements upstream to their own partners, who in turn do likewise across successive tiers. The aim is to propagate responsible practices smoothly throughout the entire network. She et al. (2025) further explored this relationship between firms and their first-tier suppliers. They found that improving this sustainability-related (carbon) disclosure helps in the improvement of collaboration propensity. Thus, institutional pressure can exert a further spillover effect on the attempt to foster transparency and sustainable supply chain practices.

However, maximizing the benefits of the directive means going beyond more regulatory compliance: Firms need to be galvanized to deeply integrate sustainability into their core business practices so that their supply chains operate sustainably both on paper and in practice. As companies increasingly look to align with global sustainability standards, it is critical to understand the barriers and reliability of data collection processes, since they will involve (and hopefully actively engage) several categories of internal and external stakeholders.

3.5 ESRS and Data Collection: Barriers and Reliability Issues

Data collection is pivotal for estimating the true impact of managerial practices within supply chains. Thus, the availability and reliability of data have long been subjects of discussion within the discipline of supply chain management. In this debate, Ellram and Tate (2016, p. 251) highlighted the advantages of using secondary data, which include the availability of relatively large datasets, cost-effectiveness, time efficiency, the potential for objectivity compared to primary data (e.g., cases or surveys), reduced personnel requirements for data collection, the ability to pool other data types for more thorough investigations, access to corporate data (which avoids reliance on historical perceptions), and the ability to use common, well-defined measures.

To contribute to the debate, this subsection discusses several expected drivers and barriers to ESRS standard implementation. As mentioned above, the new approach promoted by EFRAG requires large companies to increase the effort and resources they invest in analyzing, disclosing, and mitigating their impacts. By considering the drivers and barriers for these firms' transitions, we can highlight specific challenges related not only to the boundary conditions but also to control functions and strategic planning implications (Kazancoglu et al., 2023; Martínez-Ferrero & García-Sánchez, 2017).

With sustainable supply chain management becoming increasingly linked to sustainability performance disclosure, rather than mere operational efficiency, firms can no longer postpone the integration of a structured sustainability function at the corporate level (within the company or industrial group) (Caccialanza, 2022; Kazancoglu et al., 2023). Thus, the debate over drivers and barriers revolves around the tensions and interconnections between internal and external dynamics. Numerous scholars have already attempted to classify, group, and evaluate external drivers and barriers for the promotion of sustainable (or green) supply chain management (Akkartal & Mızrak, 2023). For instance, Govidan et al. (2014) noted that barriers can include several dimensions from financial and technological to knowledge and involvement. These aspects can further evolve, as certain barriers are being mitigated by the increasing democratization and accessibility of solutions (particularly technological and digital) that once posed significant obstacles to fully realizing their potential (Paulraj, 2011). This has resulted in a standardized set of metrics that eases data collection and systematization (see the Draft EFRAG IG 3 List of ESRS Data Points) and gives external parties more digital avenues to communicate with the company and demand accountability, thus, in turn, making the demand for (expected) accountability even greater (Gualandris et al., 2021; Mızrak & Cevher, 2023; Pizzi et al., 2021; Yu et al., 2025).

A survey conducted by the PWC (Global CSRD Survey, 2024) shows the expected extent of the determined barriers to collecting affordable yet reliable data. As Fig. 3.7 illustrates, 59% of managers are significantly worried about data availability, while 57% are very concerned about value chain complexity. Conversely, the involvement of management is perceived to be the least pressing barrier.

Increasingly, companies feel the need to engage in formal communication about their performance to legitimize themselves with their stakeholders (Bowrey & Clements, 2019; de Oliveira et al., 2018). However, this communication only has repercussions if supported by an actual commitment and endorsement from top management (Blome et al., 2014; Martínez-Ferrero & García-Sánchez, 2017). This way, by demonstrating this commitment to products' sustainability-oriented attributes, they are better equipped to defend their distinctive product positioning (Khalid & Seuring, 2019). Nevertheless, many companies are clearly reluctant to disclose their performance, especially in SME contexts and in situations where communication (over)exposes compared to the control of reported data (Galli et al., 2023). Meanwhile, factors such as organizational culture can moderate the propensity for sustainable disclosures. Specifically, culture can be a facilitative or limiting element—both in terms of (sustainability) culture at the corporate level and as a factor shaping reporting expectations at the macro level (Pizzi et al., 2022). Similarly, this cultural element can impact how well firms engage with various stakeholder categories during performance reporting—whether that relates to listening to their concerns, defining objectives, and addressing relevant issues (see Sect. 2.1). The redefinition of objectives—particularly through the integration of external

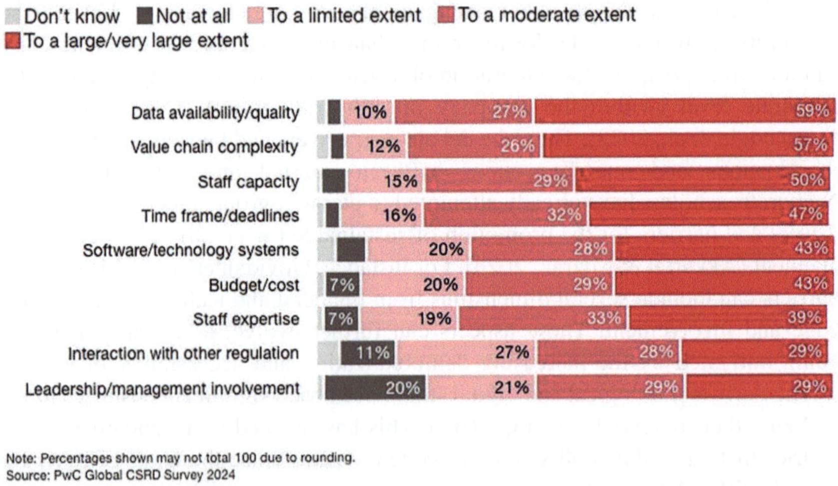

Fig. 3.7 The perceived main obstacle to CSRD implementation (n = 547). *Source*: PwC Global CSRD Survey 2024

demands—reflects a broader adjustment in how organizations interact with their context, shaped by adaptive, reactive, or proactive approaches (Beske & Seuring, 2014).

According to the EFRAG study on the implementation of ESRS, there are warning signs that firms are not meeting requirements, given the low level of readiness declared. Specifically, the EFRAG (2024, p. 6) study noted that

- "~80% of the undertakings mention the complexity of retrieving data, with similar obstacles found across E, S, and G.
- ~75% of the undertakings adopt the phase-in option, while ~40% currently use Information Materiality at a data point level, and ~ 10% target full disclosure of data points related to material topics.
- ~95% of the undertakings use Implementation Guidance 3 for running their gap analysis, while ~20% are currently using it as a preliminary base for the upcoming digital tagging/reporting taxonomy (XBRL taxonomy)."

Additional impetus arises from successful cases and best practices implemented by other firms (including direct competitors), which, in line with our framework, activate the dynamics of reactive strategic adaptation among follower companies (Akkartal & Mızrak, 2023; Balon, 2020). These elements would create further incentives for external institutional pressures, in line with prior studies (Kauppi & Hannibal, 2017; Pizzi et al., 2021; Scarpa et al., 2023; Wu et al., 2013; Zimon et al., 2020). A consistent difference exists, however, between the scope of the NFRD and that of the more recent CSRD and CSDDD. The former aimed to promote the disclosure of non-financial performance at the level of individual companies or corporate groups. The latter not only seeks to enhance the uniformity and comparability of sustainability disclosure but also extends the focus to include impacts across the value chain.

3.6 The Omnibus Package

During 2024, a structural shift took shape in the European Union's sustainability policy strategy, crystallized in two flagship strategic texts: the report The Future of European Competitiveness, prepared under Mario Draghi's leadership and released on September 9, 2024, and the European Commission's January 29, 2025, Communication, A Competitiveness Compass for the EU. While not legally binding, these documents have reframed both political and technical debates, charting a new course that seeks to balance the urgency of the green transition with the competitive needs of Europe's production systems.

Both texts highlight significant shortcomings in the EU's current sustainability rulebook—especially the Corporate Sustainability Reporting Directive (CSRD) and the European Sustainability Reporting Standards (ESRS). Chief concerns include regulatory complexity and overlaps; heavier administrative burdens, particularly for small and medium-sized enterprises (SMEs); implementation difficulties flagged by international stakeholders; and possible adverse effects on productivity and consumer prices. These pressures are compounded by a volatile global context and uncertainty around the industrial strategies of major economies such as China, India, and the United States.

Against this backdrop, on February 26, 2025, the Commission presented the "Omnibus simplification package on sustainability" (the Omnibus Package), a legislative initiative aimed at streamlining the ESG regulatory framework. The package forms part of the broader A Simpler and Faster Europe: Communication on Implementation and Simplification, which underpins the Commission's 2024–2029 strategic plan. It pursues a dual objective: (i) reduce regulatory and compliance burdens, targeting a 25% cut for large companies and 35% for SMEs; and (ii) bolster Europe's industrial competitiveness while catalyzing about €50 billion in public and private investment and generating approximately €6.3 billion in annual administrative savings.

The package comprises three proposals. Proposal 1 ("Stop-the-Clock") defers CSRD reporting obligations for companies in the second and third implementation waves ("wave 2" and "wave 3"), pushing entry into force to 2028 and 2029, respectively. It also extends the transposition deadline for the Corporate Sustainability Due Diligence Directive (CSDDD) by 1 year to July 26, 2027, with the first application from July 26, 2028. In short, Proposal 1 resets the dates from which Member States must apply specified sustainability reporting and due diligence obligations.

Proposal 2 makes more structural changes to the CSRD, the Accounting Directive, the Audit Directive, and the CSDDD. Key elements include raising the CSRD size thresholds so that the directive would apply only to companies with more than 1000 employees, thereby narrowing the scope of mandatory reporters. It also introduces a "value chain cap" to limit the trickle-down of information demands along the chain, protecting SMEs from excessive requests. Due diligence obligations would be confined to direct business partners. For the time being, the shift from limited assurance to reasonable assurance over sustainability information would be put on hold. Sector-specific ESRS would be postponed by 2 years: whereas the original CSRD (Art. 29b sexies) required adoption by June 2024, the Omnibus Proposal moved the deadline to June 30, 2026—reducing the near-term load while preserving future sectoral differentiation. In addition, Proposal 2 reduces the total number of data points in the general (cross-cutting and topical) ESRS: more than 300 items are moved from "shall disclose" to "voluntary subject to materiality" or shifted into non-binding guidance. The explicit aim is to streamline the first reporting cycle, align disclosure burdens with current data collection capabilities, and avoid duplication with forthcoming sector standards.

Proposal 3 is a delegated act revising the EU Taxonomy for sustainable investments. Notable changes include exempting activities that are financially immaterial (less than 10% of revenue, capex, or total assets) from reporting, simplifying reporting templates and the Do No Significant Harm (DNSH) criteria, and recalibrating the Green Asset Ratio (GAR) by removing exposures to companies outside the CSRD's scope from the denominator.

Beyond these three proposals, the package also amends other instruments, such as the Carbon Border Adjustment Mechanism (CBAM) and the InvestEU Regulation, though these adjustments are not directly relevant to the present analysis of adequate organizational arrangements in supply chains.

A central element of the Omnibus approach is recognition of the VSME, the voluntary sustainability reporting standard for unlisted SMEs outside the CSRD perimeter. As discussed in 4.4, the VSME is designed to ease SMEs' transition toward sustainability and provide a shared informational benchmark. Within the Omnibus logic, although not legally binding, the VSME is intended to rationalize information requests along the value chain by defining a reasonable baseline of disclosure expected from non-obligated suppliers to obligated customers. In practice, SMEs that prepare a sustainability report aligned with the VSME would be considered as meeting transparency expectations from CSRD-subject clients, reducing the risk of excessive or duplicative information demands.

As for the legislative pathway, Proposal 1 has already completed the approval process. Proposals 2 and 3 will proceed differently: Proposal 2 will be examined by the European Parliament and the Council, whereas Proposal 3, as a delegated act, may be adopted directly by the Commission unless objected to during the scrutiny period. While the Omnibus Package does not alter the conceptual foundations of EU sustainability reporting—the double-materiality principle remains the model's backbone—it aims to make the framework more efficient and proportionate through targeted simplification and rationalization, calibrated to firms' size and operational capacity.

On February 26, 2025, following unanimous approval by the Sustainability Reporting Board (SRB), EFRAG signaled its readiness to support the Commission on technical measures stemming from the Omnibus proposals and immediately launched the initial work. Subsequently, by a letter dated March 27, 2025, Commissioner Albuquerque mandated EFRAG to deliver an ESRS simplification package, setting October 31, 2025, as the deadline for technical advice and requesting a detailed work plan by April 15, 2025 (Albuquerque, 2025).

With the October 31, 2025, deadline in view for submitting the revised ESRS draft to the Commission, EFRAG established an internal timetable focusing available time on technical tasks to ensure quality and completeness. From a governance standpoint—and given the compressed schedule—EFRAG's governing bodies (the Administrative Board, SRB, and SR TEG) will closely monitor each phase within their respective remits. In narrative form, the plan (EFRAG, 2025, p. 2, ESRS Revision: Work Plan and Timeline) proceeds as follows:

- April to mid-May 2025: Define a vision and operational levers for meaningful simplification (to be validated through stakeholder feedback) and gather evidence from stakeholders and published reports.
- Late May to July 2025: Draft and approve the exposure drafts (EDs) amending the ESRS.
- August to September 2025: Publish EDs and collect/analyze stakeholder feedback, including public consultation.
- October 2025: Finalize and transmit the technical advice to the European Commission.

Looking ahead to forthcoming simplification measures, European firms face a new phase of both challenges and opportunities to disclose their sustainability performance. If implemented as designed, the Omnibus measures—together with the VSME—could reduce duplicative requests and right-size disclosure expectations along value chains and help SMEs build scalable reporting capabilities without compromising the EU's overarching sustainability objectives. Chapters 4 and 5 further explore how European regulation will practically affect sustainability performance evaluations.

References

Abbasi, M., & Nilsson, F. (2012). Themes and challenges in making supply chains environmentally sustainable. *Supply Chain Management: An International Journal, 17*(5), 517–530.

Akkartal, G. R., & Mızrak, F. (2023). Defining reverse and green supplier selection criteria with a strategic management approach. In *Economic development and the environmental ecosystem: The role of energy policy in economic growth* (pp. 205–218). Springer Nature.

Amaeshi, K. M., Osuji, O. K., & Nnodim, P. (2008). Corporate social responsibility in supply chains of global brands: A boundaryless responsibility? Clarifications, exceptions and implications. *Journal of Business Ethics, 81*(1), 223–234.

Andersson, F. N., & Arvidsson, S. (2023). The EU'S sustainable finance platform: A new game plan in the quest for competitive advantage. In *Creating a sustainable competitive position: Ethical challenges for international firms* (Vol. 37, pp. 237–249). Emerald Publishing Limited.

Antonini, C., & Gomez-Conde, J. (2024). Environmental management control systems and environmental innovation: Unintended consequences of the EU non-financial reporting directive. *Management Accounting Research, 65*, 100903.

Aureli, S., Del Baldo, M., Lombardi, R., & Nappo, F. (2020). Nonfinancial reporting regulation and challenges in sustainability disclosure and corporate governance practices. *Business Strategy and the Environment, 29*(6), 2392–2403.

Aureli, S., Magnaghi, E., & Salvatori, F. (2019). The role of existing regulation and discretion in harmonising non-financial disclosure. *Accounting in Europe, 16*(3), 290–312.

Baboukardos, D., Gaia, S., Lassou, P., & Soobaroyen, T. (2023). The multiverse of non-financial reporting regulation. *Accounting Forum, 47*(2), 147–165.

Balon, V. (2020). Green supply chain management: Pressures, practices, and performance—An integrative literature review. *Business Strategy & Development, 3*(2), 226–244.

Baltrunaite, A., Canini, R., Chiarelli, F., Consigliere, E., Coppotelli, P., Fornasari, F., Sette, E., et al. (2024). *The corporate sustainability due diligence directive (CSDDD): An analysis of the potential economic and legal impacts* (Vol. No. 869). Bank of Italy, Economic Research and International Relations Area.

Beske, P., & Seuring, S. (2014). Putting sustainability into supply chain management. *Supply Chain Management: An International Journal, 19*(3), 322–331.

Beske-Janssen, P., Johnsen, T., Constant, F., & Wieland, A. (2023). New competences enhancing procurement's contribution to innovation and sustainability. *Journal of Purchasing and Supply Management, 29*(3), 100847.

Biondi, L., Dumay, J., & Monciardini, D. (2020). Using the international integrated reporting framework to comply with EU directive 2014/95/EU: Can we afford another reporting façade? *Meditari Accountancy Research, 28*(5), 889–914.

Blome, C., Hollos, D., & Paulraj, A. (2014). Green procurement and green supplier development: Antecedents and effects on supplier performance. *International Journal of Production Research, 52*(1), 32–49.

Bowrey, G., & Clements, M. (2019). Supply chain legitimation through CSR reporting. *Australasian Accounting, Business and Finance Journal, 13*(1), 27–43.

Caccialanza, A. (2022). Supply chain management and procurement. In R. Brinkmann (Ed.), *The Palgrave handbook of global sustainability*. Palgrave Macmillan. https://doi.org/10.1007/978-3-030-38948-2_101-1

Camilleri, M. A. (2020). European environment policy for the circular economy: Implications for business and industry stakeholders. *Sustainable Development, 28*(6), 1804–1812.

CDP. (2022). *Global supply chain report 2021*. https://cdn.cdp.net/cdp-production/cms/reports/documents/000/006/106/original/CDP_SC_Report_2021.pdf?1644513297

Chelli, M., Durocher, S., & Fortin, A. (2018). Normativity in environmental reporting: A comparison of three regimes. *Journal of Business Ethics, 149*, 285–311.

Costa, E., & Agostini, M. (2016). Mandatory disclosure about environmental and employee matters in the reports of Italian-listed corporate groups. *Social and Environmental Accountability Journal, 36*(1), 10–33.

Čufar, M., Belak, J., & Primec, A. (2024). Beyond compliance: Leveraging NFRD and CSRD for transformative CSR reporting and stakeholder empowerment in corporate governance. In *Corporate governance and CSR strategies for sustainability* (pp. 233–261). IGI Global.

Cuomo, F., Gaia, S., Girardone, C., & Piserà, S. (2024). The effects of the EU non-financial reporting directive on corporate social responsibility. *The European Journal of Finance, 30*(7), 726–752.

Daddi, T., Heras-Saizarbitoria, I., Marrucci, L., Rizzi, F., & Testa, F. (2021). The effects of green supply chain management capability on the internalisation of environmental management systems and organisation performance. *Corporate Social Responsibility and Environmental Management, 28*(4), 1241–1253.

Dai, J., Xie, L., & Chu, Z. (2021b). Developing sustainable supply chain management: The interplay of institutional pressures and sustainability capabilities. *Sustainable Production and Consumption, 28*, 254–268.

Dai, R., Liang, H., & Ng, L. (2021a). Socially responsible corporate customers. *Journal of Financial Economics, 142*(2), 598–626.

Das, D. (2017). Development and validation of a scale for measuring sustainable supply chain management practices and performance. *Journal of Cleaner Production, 164*, 1344–1362.

Davis, J., & DeWitt, T. (2024, July 30). *Can the field of strategy take on the climate crisis? In Europe, yes; in the US, maybe not*. Management Studies Insights. https://managementstudiesinsights.com/can-the-field-of-strategy-take-on-the-climate-crisis-in-europe-yes-in-the-us-maybe-not/

de Oliveira, U. R., Espindola, L. S., da Silva, I. R., da Silva, I. N., & Rocha, H. M. (2018). A systematic literature review on green supply chain management: Research implications and future perspectives. *Journal of Cleaner Production, 187*, 537–561.

Diabat, A., Kannan, D., & Mathiyazhagan, K. (2014). Analysis of enablers for implementation of sustainable supply chain management—A textile case. *Journal of Cleaner Production, 83*, 391–403.

Dinh, T., Husmann, A., & Melloni, G. (2023). Corporate sustainability reporting in Europe: A scoping review. *Accounting in Europe, 20*(1), 1–29.

DIRECTIVE (EU). (2024). *2024/1760 of the European parliament and of the council of 13 June 2024 on corporate sustainability due diligence and amending Directive (EU) 2019/1937 and Regulation (EU) 2023/2859*. https://eur-lex.europa.eu/legal-content/EN/TXT/PDF/?uri=OJ:L_202401760

Dong, S., Xu, L., & McIver, R. P. (2023). Sustainability reporting quality and the financial sector: Evidence from China. *Meditari Accountancy Research, 31*(5), 1190–1214.

EFRAG (2022a). *Public consultation on the first set of Draft ESRS*. https://www.efrag.org/lab3?AspxAutoDetectCookieSupport=1#subtitle6

EFRAG. (2022b). *[Draft] EFRAG IG 3 list of ESRS data points* (Last update: 23/12/2022). https://efrag.sharefile.com/share/view/s1a12c193b86d406e90b1bcd7b6bb8f6f/fo37c90b-9d9b-4432-a76b-27760cfcc01b

EFRAG. (2023a). *[Draft] EFRAG IG 1: Materiality assessment implementation guidance*. https://www.efrag.org/Assets/Download?assetUrl=%2Fsites%2Fwebpublishing%2FSiteAssets%2FDraft%2520EFRAG%2520IG%25201%2520MAIG%2520231222.pdf
EFRAG. (2023b). *[Draft] EFRAG IG 2: Value chain implementation guidance*. https://www.efrag.org/Assets/Download?assetUrl=%2Fsites%2Fwebpublishing%2FSiteAssets%2FDraft%2520EFRAG%2520IG%25202%2520VCIG%2520231222.pdf
EFRAG. (2024). *Implementation of ESRS: Initial observed practices from selected companies*. https://www.efrag.org/sites/default/files/media/document/2024-07/EFRAG_ESRS%20initial%20observed%20practices%20Q2%202024%20final%20version.pdf
EFRAG. (2025). *[Draft] Amended ESRS 2—Exposure draft—July 2025*. European Financial Reporting Advisory Group. https://www.efrag.org/sites/default/files/media/document/2025-07/Amended_ESRS_Exposure_Draft_July_2025_ESRS_2.pdf
Ellram, L. M., & Tate, W. L. (2016). The use of secondary data in purchasing and supply management (P/SM) research. *Journal of Purchasing and Supply Management, 22*(4), 250–254.
ESG Book. (2023). *Reporting Exchange—The adoption of CSRD and ESRS*. https://www.esgbook.com/insights
European Commission. (2014). *Non-financial reporting EU rules require large companies to publish regular reports on the social and environmental impacts of their activities*. Overview. https://ec.europa.eu/info/business-economy-euro/company-reporting-and-auditing/company-reporting/non-financial-reporting_en
European Commission. (2021). *Questions and answers: Corporate sustainability reporting directive proposal*. https://ec.europa.eu/commission/presscorner/detail/en/qanda_21_1806
European Commission. (2024). *Corporate sustainability due diligence directive: Frequently asked questions*. https://commission.europa.eu/document/download/7a3e9980-5fda-4760-8f25-bc5571806033_en?filename=240719_CSDD_FAQ_final.pdf
European Financial Reporting Advisory Group. (2025, 27 March). *Letter from Commissioner Maria Luís Albuquerque mandating EFRAG to deliver ESRS simplification technical advice. EFRAG*. https://api.knowledgehub.efrag.org/dev-docattachment/commissioner-albuquerque-letter-to-efrag-march-2025.pdf
European Union. (2021). *Platform on sustainable finance*. https://finance.ec.europa.eu/sustainablefinance/overview-sustainable-finance/platform-sustainable-finance_en
Eurostat. (2025, March). *International trade in goods*. https://ec.europa.eu/eurostat/statistics-explained/index.php/International_trade_in_goods
Fan, Y., & Stevenson, M. (2018). A review of supply chain risk management: Definition, theory, and research agenda. *International Journal of Physical Distribution & Logistics Management, 48*(3), 205–230.
Fazzini, M., & Dal Maso, L. (2016). The value relevance of "assured" environmental disclosure: The Italian experience. *Sustainability Accounting, Management and Policy Journal, 7*(2), 225–245.
Formentini, M., & Taticchi, P. (2016). Corporate sustainability approaches and governance mechanisms in sustainable supply chain management. *Journal of Cleaner Production, 112*, 1920–1933.
Frank Bold. (2025a). *Revised EU sustainability reporting standards: Analysis of the changes and trade-offs*. Frank Bold.
Frank Bold. (2025b, July 31). *Our take on EFRAG's proposals for simplified EU Sustainability Reporting Standards*. Frank Bold News. https://en.frankbold.org/news/our-take-on-efrags-proposals-for-simplified-eu-sustainability-reporting-standards
Galli, D., Torelli, R., & Caccialanza, A. (2023). Sustainability performance and sustainability reporting in SMEs: A love affair or a fight? *Journal of Management & Organization*, 1–26.
Garcia-Torea, N., Luque-Vílchez, M., & Rodríguez-Gutiérrez, P. (2024). The EU taxonomy, sustainability reporting and financial institutions: Understanding the elements driving regulatory uncertainty. *Accounting Forum, 48*(3), 1–28.
Gauzès, J. P., & Board, E. (2021). *Potential need of changes to the governance and funding of EFRAG*. European Reporting Lab. https://leasenews.it/storage/app/uploads/public/604/643/d36/604643d36bcf0117093624.pdf.

Gold, S., & Wieland, A. (Eds.). (2024). *The supply chain: A system in crisis*. Edward Elgar.

GoldenBee Consulting. (2023). *Top 10 ESG events in China in 2022*. https://en.goldenbeechina.com/Home/News/show/id/212

González, M., Alonso-Almeida, M. D. M., & Dominguez, D. (2018). Mapping global sustainability report scoring: A detailed analysis of Europe and Asia. *Quality & Quantity, 52*, 1041–1055.

Govindan, K., & Bouzon, M. (2018). From a literature review to a multi-perspective framework for reverse logistics barriers and drivers. *Journal of Cleaner Production, 187*, 318–337.

Govindan, K., Kaliyan, M., Kannan, D., & Haq, A. N. (2014). Barriers analysis for green supply chain management implementation in Indian industries using analytic hierarchy process. *International Journal of Production Economics, 147*, 555–568.

Gualandris, J., Longoni, A., Luzzini, D., & Pagell, M. (2021). The association between supply chain structure and transparency: A large-scale empirical study. *Journal of Operations Management, 67*(7), 803–827.

Guo, H., Shen, Z., Chen, Y., & Dong, M. (2025). Analyzing the impact of government R&D subsidy and digital transformation on supply chain risk dynamics management and firm performance in the China's chip industry. *International Journal of Production Economics, 281*, 109524.

He, C., & Loftus, J. (2014). Does environmental reporting reflect environmental performance? Evidence from China. *Pacific Accounting Review, 26*(1/2), 134–154.

Higham, I. (2019). UN guiding principles on business and human rights. In *Standardization and risk governance* (pp. 217–234). Routledge.

Hoffmann, E., Dietsche, C., & Christine, H. (2018). Between mandatory and voluntary: Non-financial reporting by German companies. *Sustainability Management Forum, 26*(1–4), 47–63.

Huld, A. (2022). China ESG reporting—New measures on disclosure of enterprise environmental information. *China Briefing*. https://www.china-briefing.com/news/china-esg-reporting-disclosing-enterprise-environmental-information/

Hurt, J., Ledebur, K., Meyer, B., Friesenbichler, K., Gerschberger, M., Thurner, S., & Klimek, P. (2023). Supply chain due diligence risk assessment for the EU: A network approach to estimate expected effectiveness of the planned EU directive. *arXiv preprint arXiv*, 2311.15971.

ILO. (2022). *Tripartite declaration of principles concerning multinational enterprises and social policy*. https://www.ilo.org/wcmsp5/groups/public/%2D%2D-ed_emp/%2D%2D-emp_ent/%2D%2D-multi/documents/publication/wcms_094386.pdf

International Monetary Fund - IMF. (2025, April). *World Economic Outlook database: GDP based on PPP share of world; GDP, current prices*. https://www.imf.org/external/datamapper/PPPSH@WEO/OEMDC/ADVEC.

Ioannou, I., & Serafeim, G. (2017). *The consequences of mandatory corporate sustainability reporting* (pp. 11–100). Harvard Business School Research Working Paper.

Isaksson, O. H., Simeth, M., & Seifert, R. W. (2016). Knowledge spillovers in the supply chain: Evidence from the high tech sectors. *Research Policy, 45*(3), 699–706.

Ivanov, D. (2018). Revealing interfaces of supply chain resilience and sustainability: A simulation study. *International Journal of Production Research, 56*(10), 3507–3523.

Jensen, J. C., & Berg, N. (2012). Determinants of traditional sustainability reporting versus integrated reporting. An institutionalist approach. *Business Strategy and the Environment, 21*(5), 299–316.

John, R., Jaeger-Erben, M., & Rückert-John, J. (2016). Elusive practices: Considerations on limits and possibilities of environmental policy for sustainable consumption. *Environmental Policy and Governance, 26*(2), 129–140.

Junior, R. M., Best, P. J., & Cotter, J. (2014). Sustainability reporting and assurance: A historical analysis on a world-wide phenomenon. *Journal of Business Ethics, 120*, 1–11.

Kauppi, K., & Hannibal, C. (2017). Institutional pressures and sustainability assessment in supply chains. *Supply Chain Management: An International Journal, 22*(5), 458–472.

Kazancoglu, I., Sagnak, M., Kumar Mangla, S., & Kazancoglu, Y. (2021). Circular economy and the policy: A framework for improving the corporate environmental management in supply chains. *Business Strategy and the Environment, 30*(1), 590–608.

Kazancoglu, Y., Ozkan-Ozen, Y. D., Sagnak, M., Kazancoglu, I., & Dora, M. (2023). Framework for a sustainable supply chain to overcome risks in transition to a circular economy through industry 4.0. *Production Planning & Control, 34*(10), 902–917.

Khalid, R. U., & Seuring, S. (2019). Analyzing base-of-the-pyramid research from a (sustainable) supply chain perspective. *Journal of Business Ethics, 155*, 663–686.

Khalid, R. U., Seuring, S., Beske, P., Land, A., Yawar, S. A., & Wagner, R. (2015). Putting sustainable supply chain management into base of the pyramid research. *Supply Chain Management: An International Journal, 20*(6), 681–696.

Koberg, E., & Longoni, A. (2019). A systematic review of sustainable supply chain management in global supply chains. *Journal of Cleaner Production, 207*, 1084–1098.

KPMG China. (2024). *China stock exchanges finalised mandatory sustainability reporting requirements for larger listed entities*. https://assets.kpmg.com/content/dam/kpmg/cn/pdf/en/2024/04/china-stock-exchanges-finalised-mandatory-sustainability-reporting-requirements-for-larger-listed-entities.pdf

La Torre, M., Sabelfeld, S., Blomkvist, M., Tarquinio, L., & Dumay, J. (2018). Harmonising non-financial reporting regulation in Europe: Practical forces and projections for future research. *Meditari Accountancy Research, 26*(4), 598–621.

Le, T. T., Vo, X. V., & Venkatesh, V. G. (2022). Role of green innovation and supply chain management in driving sustainable corporate performance. *Journal of Cleaner Production, 374*, 133875.

Lieberman, M. B., & Asaba, S. (2006). Why do firms imitate each other? *Academy of Management Review, 31*(2), 366–385.

Lykkesfeldt, P., & Kjaergaard, L. L. (2022). *Investor relations and ESG reporting in a regulatory perspective: A practical guide for financial market participants*. Springer Nature.

Maas, K., & Sampers, P. (2020). The expected impacts of regulating non-financial reporting. *Maandblad Voor Accountancy en Bedrijfseconomie, 94*(7/8), 265–274.

Maas, K. E. H., & Vermeulen, M. C. (2015). *A systemic view on the impacts of regulating non-financial reporting*. Erasmus School of Economics.

Mak, C. (2022). Corporate sustainability due diligence: More than ticking the boxes? *Maastricht Journal of European and Comparative Law, 29*(3), 301–303.

Martínez-Ferrero, J., & García-Sánchez, I. M. (2017). Coercive, normative and mimetic isomorphism as determinants of the voluntary assurance of sustainability reports. *International Business Review, 26*(1), 102–118.

Martin-Ortega, O. (2022). Transparency and human rights in global supply chains: From corporate-led disclosure to a right to know. In *Research handbook on global governance, business and human rights* (pp. 100–120). Edward Elgar.

Miandar, T., Johnsen, E. T., & Caniato, F. (2023). The role of purchasing and supply management in diffusing sustainability in supply networks: A systematic literature review. *Business Ethics, the Environment & Responsibility, 33*, 505.

Mızrak, F., & Cevher, M. F. (2023). The impact of social marketing and corporate social responsibility on energy savings as a competitive strategy. In *Handbook of research on achieving sustainable development goals with sustainable marketing* (pp. 239–254). IGI Global.

Moggi, S., Lehman, G., & Pagani, A. (2023). The juridification of social accounting and the transposition process of the non-financial reporting directive 2014/95/EU. *Meditari Accountancy Research, 31*(7), 185–208.

Monciardini, D., Mähönen, J. T., & Tsagas, G. (2020). Rethinking non-financial reporting: A blueprint for structural regulatory changes. *Accounting, Economics, and Law: A Convivium, 10*(2), 20200092.

Morali, O., & Searcy, C. (2013). A review of sustainable supply chain management practices in Canada. *Journal of Business Ethics, 117*, 635–658.

Muserra, A. L., Papa, M., & Grimaldi, F. (2020). Sustainable development and the European Union policy on non-financial information: An Italian empirical analysis. *Corporate Social Responsibility and Environmental Management, 27*(1), 22–31.

Odobaša, R., & Marošević, K. (2023). Expected contributions of the European corporate sustainability reporting directive (CSRD) to the sustainable development of the European union. *EU and comparative law issues and challenges series (ECLIC), 7*, 593–612.
OECD. (2023). *Guidelines for multinational enterprises*. https://www.oecd.org/corporate/mne/
Olsen, M., & Boxenbaum, E. (2009). Bottom-of-the-pyramid: Organizational barriers to implementation. *California Management Review, 51*(4), 100–125.
Pagani, A., Moggi, S., & Gaetano, A. (2021). Non-financial disclosure assurance: Critical insights from the Italian financial services sector. *International Journal of Business and Management, 16*(1), 8–18.
Parrondo, L. (2024). EFRAG roadmap for new developments in ERS reporting. In *Research handbook on financial accounting* (pp. 138–155). Edward Elgar Publishing.
Paulraj, A. (2011). Understanding the relationships between internal resources and capabilities, sustainable supply management and organizational sustainability. *Journal of Supply Chain Management, 47*(1), 19–37.
Pedersen, E. R. G., & Andersen, K. R. (2023). Greenwashing: A broken business model. *Journal of Business Models, 11*(2), 11–24.
Pickering, J., Bäckstrand, K., & Schlosberg, D. (2020). Between environmental and ecological democracy: Theory and practice at the democracy-environment nexus. *Journal of Environmental Policy & Planning, 22*(1), 1–15.
Pizzi, S., Del Baldo, M., Caputo, F., & Venturelli, A. (2022). Voluntary disclosure of sustainable development goals in mandatory non-financial reports: The moderating role of cultural dimension. *Journal of International Financial Management & Accounting, 33*(1), 83–106.
Pizzi, S., Moggi, S., Caputo, F., & Rosato, P. (2021). Social media as stakeholder engagement tool: CSR communication failure in the oil and gas sector. *Corporate Social Responsibility and Environmental Management, 28*(2), 849–859.
PwC. (2024). *PwC's global CSRD survey 2024: The promise and reality of CSRD reporting*. https://www.pwc.com/gx/en/issues/esg/global-csrd-survey.html?mod=djemSustainableBusinessPro
Rasche, A., & Waddock, S. (2021). The UN guiding principles on business and human rights: Implications for corporate social responsibility research. *Business and Human Rights Journal, 6*(2), 227–240.
Raucci, D., & Tarquinio, L. (2020). Sustainability performance indicators and non-financial information reporting. Evidence from the Italian case. *Administrative Sciences, 10*(1), 13.
Rogerson, M., Scarpa, F., & Snelson-Powell, A. (2024). Accounting for human rights: Evidence of due diligence in EU-listed firms' reporting. *Critical Perspectives on Accounting, 99*, 102716.
Ropes & Gray. (2024). *CSRD transposition tracker (monthly updated, last update 21 July 2024)*. https://www.ropesgray.com/en/insights/viewpoints/102je23/ropes-gray-publishes-eu-csrd-transposition-tracker-through-july-6-2024-subst
Sarkis, J., Hervani, A. A., & Helms, M. (2016). Applying economic non-market valuation for sustainable supply chain performance measurement and evaluation. In *Sustainable value chain management* (pp. 197–219). Routledge.
Scarpa, F., Signori, S., & Crane, A. (2025). Corporate tax responsibility: Expectations of implicit and explicit CSR in the U.K. Media. *Business & Society, 64*(2), 299–338. https://doi.org/10.1177/00076503241254826
Scarpa, F., Torelli, R., & Fiandrino, S. (2023). Business engagement for the SDGs in COVID-19 time: An Italian perspective. *Sustainability Accounting, Management and Policy Journal, 14*(7), 152–178.
Seuring, S., & Müller, M. (2008). From a literature review to a conceptual framework for sustainable supply chain management. *Journal of Cleaner Production, 16*(15), 1699–1710.
Shen, H., Wu, H., & Zhou, Y. (2025). The influence of a globalized supply chain on corporate carbon management and disclosure. *The British Accounting Review, 101696*, 101696.
Simoni, L., Bini, L., & Bellucci, M. (2020). Effects of social, environmental, and institutional factors on sustainability report assurance: Evidence from European countries. *Meditari Accountancy Research, 28*(6), 1059–1087.

Song, S., Lian, J., Skowronski, K., & Yan, T. (2024). Customer base environmental disclosure and supplier greenhouse gas emissions: A signaling theory perspective. *Journal of Operations Management, 70*(3), 355–380.
Stefanescu, C. A. (2022). Linking sustainability and non-financial reporting directive 2014/95/EU through isomorphism lens. *Meditari Accountancy Research, 30*(6), 1680–1704.
Sustainability Infographics. (2024). Positioning and structure of ESRS in the European Union legislative context (last access 28/01/2025). https://www.linkedin.com/posts/sustainability-infographics_efrag-efrag-esrs-activity-7289152588904562688-SFnk?utm_source=share&utm_medium=member_desktop
Tang, J., Wang, X., & Liu, Q. (2023). The spillover effect of customers' ESG to suppliers. *Pacific-Basin Finance Journal, 78*, 101947.
Tate, W. L., Ellram, L. M., & Kirchoff, J. F. (2010). Corporate social responsibility reports: A thematic analysis related to supply chain management. *Journal of Supply Chain Management, 46*(1), 19–44.
Torelli, R., Fiandrino, S., & Scarpa, F. (2025). Value-enhancing drivers of corporate governance in improving human rights due diligence: Worldwide evidence. *Corporate Social Responsibility and Environmental Management, 32*(1), 1279–1290.
Traxler, A. A., Greiling, D., Freinbichler, M., & Mayerhofer, P. (2023). Does the directive of the European Union on disclosure of non-financial and diversity information improve reporting? *Journal of Accounting & Organizational Change, 19*(6), 70–90.
Troisi, R., De Simone, S., & Franco, M. (2024). Illegal firm behaviour and environmental hazard: The case of waste disposal. *European Management Review, 21*(3), 605–617.
Tsai, F. M., Bui, T. D., Tseng, M. L., Ali, M. H., Lim, M. K., & Chiu, A. S. (2021). Sustainable supply chain management trends in world regions: A data-driven analysis. *Resources, Conservation and Recycling, 167*, 105421.
U.S. Census Bureau. (2025a, May). *Trade in goods with European Union, 2024 totals*. https://www.census.gov/foreign-trade/balance/c0003.html
U.S. Census Bureau. (2025b, May). *Trade in goods with China, 2024 totals*. https://www.census.gov/foreign-trade/balance/c5700.html
Veltri, S. (2020). *Mandatory non-financial risk-related disclosure*. Springer International.
Villena, V. H., & Gioia, D. A. (2020). A more sustainable supply chain. *Harvard Business Review, 98*(2), 84–93.
Villiers, C. (2022). New directions in the European Union's regulatory framework for corporate reporting, due diligence and accountability: The challenge of complexity. *European Journal of Risk Regulation, 13*(4), 548–566.
Wall Street Journal. (2025, April 9). *The U.S. and China are sti ll in a trade war. Here's how much business they do*. https://www.wsj.com/economy/trade/the-u-s-and-china-are-still-in-a-trade-war-heres-how-much-business-they-do-41d1f321
WB and IFC. (2024). *The Evolution of Sustainability Disclosure: Comparing the 2024 SEC, ESRS, and ISSB Climate Disclosures*. https://www.ifcbeyondthebalancesheet.org/understanding-global-reporting-frameworks
Wieland, A. (2021). Dancing the supply chain: Toward transformative supply chain management. *Journal of Supply Chain Management, 57*(1), 58–73.
Wu, T., Daniel, E. M., Hinton, M., & Quintas, P. (2013). Isomorphic mechanisms in manufacturing supply chains: A comparison of indigenous Chinese firms and foreign-owned MNCs. *Supply Chain Management: An International Journal, 18*(2), 161–177.
Yang, Y., Xiao, Z., Gong, Y., & al Humdan, E. (2024). The tone of buyer firms' annual reports and suppliers' green innovation: The spillover effects in the supply chain. *Business Strategy and the Environment, 33*(6), 5721–5735.
Yu, J., Yuan, G., Liu, S., & Zhang, J. (2025). Green credit, supply chain concentration and corporate digital transformation: Evidence from China. *The British Accounting Review*, 101690.
Zimon, D., Tyan, J., & Sroufe, R. (2020). Drivers of sustainable supply chain management: Practices to alignment with unsustainable development goals. *International Journal for Quality Research, 14*(1), 219–236.

Chapter 4
Evaluation and Disclosure of Sustainability Performance: From a Micro-Level to a Meso-Level Perspective

4.1 Sustainability Evaluation and Metrics

The evaluation of firms' sustainability has become a central topic in both academia and industry. Thus, there are now a large number of metrics and frameworks that can be used to measure corporate sustainability performance (Adams, 2004; Adams & McNicholas, 2007). This section explores the evolution of these metrics, providing insights into the methodologies and challenges faced in assessing sustainability (Govindan et al., 2023; Lee et al., 2024; Mohammed et al., 2018; Qorri et al., 2022; Saha & Khan, 2024).

Since the late 1990s, scholars have worked to develop systematic measurements of corporate environmental performance. For example, Ilinitch et al. (1998) noted that traditional financial metrics are insufficient for capturing the full scope of a firm's environmental impact. The relative complexity of this (non-financial) information leads to mistrust and a lack of involvement of external stakeholders (p. 385). Scholars have recognized the need for comprehensive sustainability metrics that could be integrated into the overall assessment of corporate performance. With sustainability becoming a key source of corporate competitive advantage, there is great value in identifying measures of business efficiency (Hunt & Morgan, 1996; Wang et al., 2023a). In line with this reasoning, Al-Mawali (2023) recently proposed a strategy map for various industries that draws from a sustainability balanced scorecard and decision-making trial and evaluation laboratory for manufacturing companies. This approach links sustainability objectives with strategic goals to ensure that sustainability is embedded in the core business strategy (e.g., Liu et al., 2023b).

Of course, the progression of these metrics has not gone without criticism. Dragomir (2018) pointed out several inconsistencies and challenges in existing metrics, such as the lack of standardization and the varying scope of indicators used by different studies. More specifically, Dragomir (2018) emphasized the need for a more unified and standardized approach to sustainability metrics to ensure comparability and reliability

A. Caccialanza, *Sustainability and Accountability of Supply Chains*, SIDREA Series in Accounting and Business Administration,
https://doi.org/10.1007/978-3-032-16879-5_4

across different contexts (see Cagno et al., 2023 and their proposed framework). Delmas and Blass (2010) addressed the trade-offs involved in sustainability ratings, noting that while they provide valuable benchmarks, they often suffer from methodological limitations and can lead to unintended consequences such as greenwashing. They stressed that transparency and rigor in the development of sustainability ratings are essential for enhancing their credibility and effectiveness (p. 248). Beyond consolidating multiple indicators into a single metric, scholars have suggested using a limited set of specialized indicators or progressively integrating a specific dimension (e.g., social or environmental) into corporate (financial) performance evaluation. Otherwise, sustainability-related metrics could involve the integration of issues linked to the economic-financial dimension in a more rigorous and reliable manner.

The literature suggests that integrating the social dimension into sustainability metrics will yield more results than integrating either the environmental or economic dimensions alone (Allen et al., 2009; Castellano et al., 2025; Chung 2012). A holistic sustainability assessment links environmental, social, and economic information, allowing stakeholders to see how impacts interact in value creation and risk—something that single-pillar environmental or mostly social reports cannot capture (Rinaldi et al., 2018). Research mapping the sustainability-reporting field shows that users (investors, regulators, and civil society) increasingly demand integrated disclosures because decisions about capital allocation, licenses to operate, and stakeholder trust depend on understanding cross-dimension effects, not standalone eco or social metrics (Benameur et al., 2024). Empirical evidence indicates that firms embedding broad sustainability practices—covering material social and environmental issues and linking them to governance and financial processes—outperform their peers over the long term, suggesting that multidimensional metrics drive better strategic alignment and resilience (Eccles et al., 2014; Khan et al., 2016a; Tsao et al., 2024). Triple-bottom-line perspectives reinforce this: integrating people, planet, and profit indicators reveals complementarities and trade-offs that management can exploit or mitigate, thus improving both sustainability outcomes and business performance (Nogueira et al., 2025). Taken together, contemporary sustainability-reporting scholarship points to the superiority of an integrated, multi-capital view for decision usefulness, comparability, and long-run value relative to approaches that emphasize environmental or social metrics in isolation (Benameur et al., 2024; Eccles et al., 2014; Khan et al., 2016a; Nogueira et al., 2025; Rinaldi et al., 2018). This trend is visible both in the evolution of sustainability metrics and measurement systems and in the types of firms that orient their purpose primarily toward the social or environmental pillars (Gu et al., 2024; Guo et al., 2025). To illustrate, the two pillars with reference to exemplificative examples will be discussed; at first, within the social pillar, the analysis of social enterprises is proposed, while lately, for the environmental one, the focus has been on the circular economy logics (Ecer & Torkayesh, 2022).

First, as anticipated, this evolution is clearly reflected in the social dimension of the rise of social enterprises, although there is no settled agreement in the literature on how social entrepreneurship types are grouped and what characterizes them (Conway Dato-on & Kalakay, 2016; Hota, 2023). Scholars use a spectrum of definitions for these enterprises, ranging from broad to narrow (Austin, et al., 2012;

Peredo & McLean, 2006; Galera & Borzaga, 2009). At the broad end, social entrepreneurship is treated as a mindset that can arise in any context: commercial firms, nonprofits, the public sector, or hybrids that mix for-profit and nonprofit logics (Roberts & Woods, 2005). Motivations under this broad view range from deepening an explicit social mission (common in nonprofits) to making visible social commitments in response to stakeholder pressure (common in for-profit businesses) (Conway Dato-on & Kalakay, 2016) and legitimization purposes (Bruder, 2021). The exact priority of an enterprise's social goals relative to profit does not affect its inclusion in that category (Peredo & McLean, 2006). Peredo and McLean (2006) mapped a continuum: At one pole are actors wholly motivated by producing social benefits, even without a formal enterprise structure; at the other pole are ventures that acknowledge social aims but clearly subordinate them to profit, with varying blends in between. A narrower or earned-income school of thought places social entrepreneurship squarely in the nonprofit domain, defining it as the use of commercial revenue strategies by nonprofits to fund and scale a core social mission (Saebi et al., 2019; Young & Lecy, 2014).

Across approaches, many authors converge on problem-solving and the creation and maintenance of social value as distinguishing marks of social entrepreneurship (Dees, 1998). Social and commercial entrepreneurship are better conceived as points along a continuum rather than as a binary divide, and viable initiatives must combine social relevance with economic sustainability to varying degrees (Borzaga & Tortia, 2006). Current scholarship treats social and commercial logics as the ends of a continuum navigated by hybrid organizations that must continually balance mission integrity with revenue generation. Managing the tension between social relevance and economic sustainability is now seen as a defining managerial challenge in social enterprises (Battilana & Lee, 2014; Doherty et al., 2014; Kamaludin et al., 2024; Young & Lecy, 2014). On the performance assessment side of this social dimension, Costa and Andreaus (2021) underlined the importance of stakeholder engagement in developing tailored metrics for social enterprises. They argued that participatory approaches ensure that the metrics are relevant and actionable, aligning with the unique context and goals of social enterprises. This approach aligns with efforts to create metrics that accurately reflect the social dimension of sustainability (Abbasi, 2017), which has been identified as a concern in the more exposed social enterprises (Santos, 2012; Rawhouser et al., 2017). In this vein, De Nito et al. (2024) grouped several managerial studies around these key topics (e.g., stakeholder engagement) and assessed the social impact of corporate- and project-level outcomes. In contrast, other scholars have focused not only on the dimensions of sustainability performance but also on specific categories of (purpose-driven) firms. For example, social enterprises often face challenges in maintaining their social missions while pursuing financial sustainability—a phenomenon referred to as impact drift. In their exploration of this issue, Argiolas et al. (2024) found that robust impact measurement systems can help social enterprises stay true to their core values while adapting to changing market conditions. Future work on social impact assessment can extend this analysis beyond the firms in our current sample to 'hybrid organizations' in general (Cornelissen et al., 2021; White et al., 2021), Benefit Corporations and Certified

B-Corps (Logue et al., 2025), or nonprofit organizations (Curran & Ozcan, 2024). In those cases, the expectations are recalibrated to reflect the distinctive legal duties, mission locks, third-party standards, and stakeholder accountability mechanisms embedded in these forms (Hiller, 2013; Logue et al., 2025). Because reporting obligations and assurance pathways differ—that is, statutory benefit reporting for Benefit Corps, B Lab verification and impact scoring for Certified B Corps, and donor and grantor accountability for nonprofits—further comparative frameworks will need to align impact metrics, materiality thresholds, and transparency expectations when interpreting social outcomes across these organizational statuses (for a theoretical framework, see the special issue of Ciambotti et al., 2025).

Second, other scholars have adopted a different tack: conceptualizing social enterprises as experimental arenas for assessing corporate social impacts, while simultaneously employing circular development models centered on environmental dynamics as illustrative frameworks for scaling, thereby further extending the boundaries of sustainability metrics (Amaeshi et al., 2008). In line with this reasoning, Arjaliès et al. (2023) argued for new accounting methodologies that better reflect the principles of circularity, such as resource efficiency and lifecycle assessment. They stated that traditional accounting frameworks are often inadequate for capturing the dynamics of circular economy practices and thus called for more innovative approaches. Granted, there is an ongoing debate over whether one can effectively measure the impact of circularity dynamics. Several scholars have pointed out criticisms either of the fact that firms often limit themselves to the stereotypical uses of accounting instead of expanding the paradigm (Halari & Baric, 2023) or of the effort to measure circularity that risks being too formal and thus insubstantial (Bekier & Parisi, 2023). Even more critical is the positioning of Larrinaga and Garcia-Torea (2022), who warned of the short-term perspective adopted in the conceptualization and of the risk of believing that it is possible to decouple socio-economic and biophysical systems (p. 8).

Consequently, from the perspective of evolution, scholars are advancing toward more mature analyses and comprehensive frameworks. On this topic, Ahmad et al. (2019) conducted a literature survey on sustainability indicators within the manufacturing sectors, highlighting the maturity of these indicators from a triple-bottom-line perspective (i.e., economic, environmental, and social dimensions). Their work underscored the need for continuous refinement and adaptation, which presents difficulties in finding a unique yet flexible framework that can adapt to moving targets. Nonetheless, there have been attempts: Pryshlakivsky and Searcy (2017), for instance, proposed a model to establish trade-offs in corporate sustainability performance measurement systems. They detailed a systematic approach to balancing different sustainability dimensions, ensuring that firms can make informed decisions that align their strategic goals within a cohesive framework.

On that note, the study by Vinante et al. (2021) offers hypothetical metrics for measuring circularity that are worth highlighting. According to the authors, the literature on circular economy at the firm level lacks clarity on assessment criteria: For instance, circularity indicators are not often linked to organizational functions. They analyzed 365 circularity metrics from 130 sources, arranging them into a new circular value chain framework. They highlighted the general applicability of these metrics across various firm contexts, thus facilitating comprehensive circular economy assessments. The framework reported in Fig. 4.1 reveals the complexity of reorienting internal operations and functions around sustainability while also linking logistics and (more broadly) supply chain operations with sustainability performance assessment and auditing at the supply chain level.

The model above can be extended to firms that are not formally obliged to report their performance but are indirectly required to adopt those measurement, assessment, and disclosure procedures as part of the procurement or supply chain of major European focal firms.

Beyond the aspects listed previously, technological advancements have been critical to the evolution of sustainability metrics, especially in terms of affordability and diffusion. Within this debate, Kouhizadeh et al. (2023) explored the potential of blockchain technology in enhancing circular economy performance measurements. They argued that blockchain can provide greater transparency and traceability in supply chains, thereby improving the reliability of sustainability metrics. Moreover, Casati et al. (2024) noted that the benefits of blockchain technology in supply chains are maximized when companies engage stakeholders in the integration and leverage trust among economic agents. Although blockchain technology is not yet seen as a strategic resource, its potential to enhance operational, adaptive, absorptive, and innovative capabilities will become more evident when all actors in the supply chain understand its functionality and value. Blockchain exemplifies the potential synergies between technology and sustainability evaluation, but there are other advancements (e.g., on the interoperability of technology standards, data collection, and aggregation processes) worthy of further research to fully understand the potential of this (technological) shift.

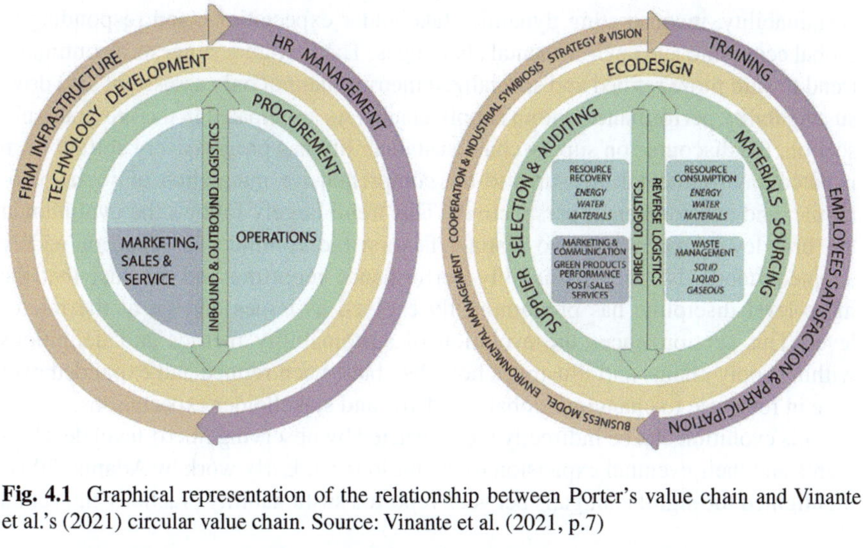

Fig. 4.1 Graphical representation of the relationship between Porter's value chain and Vinante et al.'s (2021) circular value chain. Source: Vinante et al. (2021, p.7)

In fact, all the above studies reflect a growing call for more holistic and integrated approaches to sustainability metrics (Ahi & Searcy, 2015). Ahmad et al. (2023) emphasized the need for more integrated and holistic approaches to sustainability that encompass economic, environmental, and social aspects. They noted the importance of rethinking traditional business models to better align them with long-term sustainability principles (Blanco, 2021; Cantele et al., 2020; Moggi & Dameri, 2021), specifically circular life cycle perspectives (Aureli et al., 2023).

The literature review above makes clear that corporate sustainability evaluations have evolved significantly. While early efforts focused on developing specific environmental performance indicators, later research has sought to integrate the social and economic dimensions. This shift has encouraged developments—such as sustainability ratings, heuristic models, and advanced technologies like blockchain—that have enriched the field. However, there are still challenges in achieving standardization and ensuring the reliability of sustainability metrics. Thus, there are high expectations for the new European standards for reporting sustainability performance with the ESRS (for a deeper discussion, see Sect. 4.3, particularly Sect. 3.3.3). As the field continues to become mainstream, scholars will need to uncover holistic approaches that can effectively capture the complexities of corporate sustainability performance, including not only firms' public policy positioning (Lyon et al., 2018) but also the systemic impact of business activities beyond the company's (reporting) perimeter. The next section discusses these issues in detail.

4.2 Defining the Sustainability Performance of Supply Chains

This evolution has been driven by an expanding understanding of what constitutes sustainability, incorporating dynamic stakeholder expectations, and responding to global economic and environmental challenges. The literature suggests a continuing trend toward more refined and specialized metrics that not only assess but also drive sustainability performance within supply chains. As anticipated in the previous subsection, the discourse on supply chain sustainability has progressively shifted from isolated environmental measures to the comprehensive integration of social, economic, and environmental perspectives. This trend largely follows the evolution at the firm level, except for two distinct factors: the continued absence of widely adopted standards for sustainability performance reporting and the fact that the meso-level discipline has predominantly evolved as issues emerge at the micro-level. This section traces the evolution of sustainability metrics and definitions within supply chains and illustrates how they have been refined and expanded over time in response to changing global standards and stakeholder expectations.

This evolution can be indirectly reconstructed by observing micro-level developments and their eventual expansion to the meso level. Early work by Adams (2004) highlighted the significant gaps between reported sustainability practices and actual performance, illustrating the need for more reliable and meaningful metrics. Later, Adams and McNicholas (2007) emphasized the transformative role of sustainability reporting in driving organizational change. Together, these studies suggest an initial phase in which environmental considerations dominated the discourse, leading to a primary emphasis on compliance and reporting standards at the firm (micro) level.

As the field matured and the literature on sustainability reporting deepened, the focus expanded beyond mere compliance to strategic integration within core business operations. Appannan et al. (2023) contributed to this dialogue by examining how environmental management accounting can facilitate sustainable development, while Dal Maso et al. (2024) demonstrated that higher environmental performance is associated with higher valuation (specifically market-to-book value). Their findings encourage proactive incorporation of environmental metrics into business strategies, thus shifting sustainability from a peripheral to a central business concern. Cho and Patten (2007) also recognized that environmental disclosures have strategic value not just for compliance but also for legitimizing business operations in the eyes of stakeholders. Cho and Patten (2007) analysis focused on performance disclosure as a means of legitimizing economic activities by improving accountability in internal processes in response to external demands from other supply chain actors. This interpretation aligns with the analysis of reporting drivers and barriers discussed in Sect. 2.1.

Meanwhile, the adoption of circular economy principles marked a significant shift in the discourse on sustainability reporting at the meso level. Specifically, this literature has moved toward methodological innovations and the development of specialized performance measurement systems. Recently, Liao and Wen (2023) discussed the importance of considering decision-makers' attitudinal characteristics in performance evaluations, thus adding to traditional metrics that often overlook subjective elements. Similarly, Vegter et al. (2023) proposed a performance measurement system tailored for circular supply chain management. Bai et al. (2024) introduced a fuzzy group decision approach to enhance supplier selection processes, underscoring the need for sophisticated decision-making tools that align with circular economy objectives. Concurrently, Aranda-Usón et al. (2024) discussed the development of "circular accounting," which integrates material flows and dynamic capabilities into accounting practices—a novel perspective on managing resources within supply chains. Their work supplements that of Gualandris et al. (2015), who emphasized the importance of aligning accountability mechanisms with stakeholder expectations to ensure that sustainability metrics are verifiable and aligned with broader business objectives. Giannakis et al. (2020) went further by integrating complex criteria (spanning the environmental, social, and economic dimensions) with an analytic network process to evaluate supplier sustainability performance. Govindan et al. (2023, p. 4430) introduced a conceptual model depicted in Fig. 4.2 to explore sustainable partnerships between manufacturers and suppliers.

This model includes four key components: the supplier's potential and appeal for sustainable collaboration, the collaborative efforts between manufacturer and supplier, the supplier's performance, and the key performance indicators (KPIs) for evaluating potential and appeal. These elements are interconnected: for instance, the

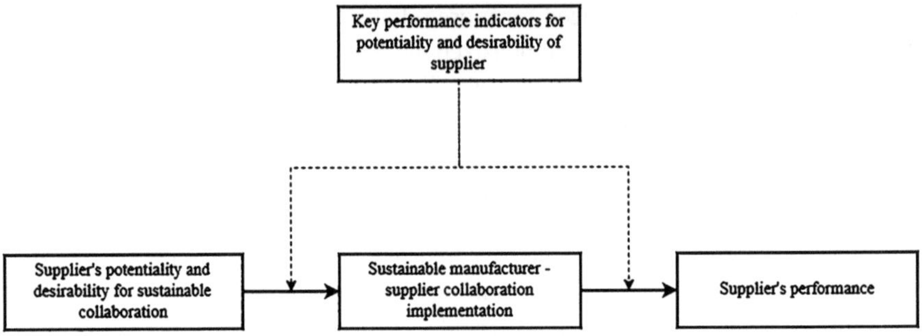

Fig. 4.2 Govindan et al. (2023) theoretical framework for sustainable manufacturer-supplier collaboration. Source: Govindan et al. (2023, p. 4430)

supplier's potential and appeal enable the success of sustainable collaboration. Consequently, performance metrics are essential for assessing the supplier's ability and willingness to engage in sustainable collaboration.

Thus, from just a purely conceptual standpoint, it is necessary to identify a tool that can position—and ideally, measure—the desirability of a potential supplier. In this regard, scholars have formulated a matrix for positioning suppliers (or their suppliers) in a classification grid based on the desirability and potentiality of the supplier (Govindan et al., 2023, p. 4440). In addition, Govindan et al. (2023, p. 4433) performed an exhaustive systematization of KPI for each of the two dimensions:

> KPIs for potentiality metric (K^p) are as follows: cost, quality, delivery, technical capability, resource consumption, financial position, innovative-ness, geographical location, flexibility, technological compatibility, facility and infrastructure, reputation, and position in the industry. The KPIs for desirability metric (K^d) are as follows: pollution prevention, waste material treatment capability, environmental competencies, training education and community influence, interests and rights of employees, environmental management system, returns handling capabilities, recycling, information disclosure, employment practices, pollution controls, green image, environmental costs, occupational health and safety management system, as well as work health and safety.

In summary, the KPIs for the supplier's potential and appeal are vital in transforming sustainable collaboration practices into measurable supplier performance outcomes.

Beyond supplier positioning, natural resource scarcity also has critical implications for supply chain sustainability strategies. Kalaitzi et al. (2018) explored how supply chains must adapt their strategies in response to resource scarcity, which sustainability initiatives seek to address. Kumar et al. (2022) extended this idea by demonstrating how sustainability priorities can be leveraged for competitive advantage. Consequently, there has been a shift in both scholarship and practice: from simple environmental measures used mainly for compliance to comprehensive disclosures that fit into an integrated strategy (Chiu et al., 2019). With the greater availability and traceability of information along the supply chain, corporate governance can more proactively assess opportunities related to supply chain configurations and dynamics. This leads to a deeper engagement process with actors who exhibit a greater orientation toward disclosure and performance reporting at the meso level, which further influences evaluation at the micro level.

4.3 Supply Chain Governance for Sustainability and the Role of Top Management Support

The topic of governance, primarily corporate governance, is a useful starting point for broadening the discourse on supply chain governance mechanisms. Here, we retrace the main aspects involved in governance issues at the corporate and supply chain levels (Gimenez & Tachizawa, 2012), specifically in light of supply chains' globalization (Gold & Wieland, 2024) and the major turn toward sustainability (Asiaei et al., 2022). This subsection also addresses the role of the board and the CSR committee in this orientation.

It is commonly accepted in the literature that good governance can boost firms' competitiveness (Claessens, 2006) and thereby ensure better operating conditions in various respects, including a lower cost of capital, higher returns on equity, greater efficiency, and greater endorsement of all stakeholders. Ideally, all stakeholders would be considered, but there has been substantial debate about prioritizing stakeholders (Letza et al., 2004; Parent & Deephouse, 2007).

Scholars have proposed different reference models for governance, rooted in differing (geographical) contexts (Bonatto et al., 2022) and cultural orientations (Jia et al., 2020), which have led to different outcomes and 'best practices' in the literature (Aguilera & Jackson, 2010). For example, Rasche (2024) recently suggested ways to implement this orientation within the analysis of three constitutive elements: the structure of boards, the mindset (of board members) involved, and the competencies of the board members. Given the numerous models and dimensions analyzed in this literature, we will predominantly focus on two aspects and dynamics that can significantly impact a model's development: the orientation toward sustainability and the globalization of supply chains.

While firms' primary approach to stakeholder management is simply optimizing their utility to key shareholders in order to maximize profits, this debate introduces a key element related to corporate social responsibility. Taking a medium- to long-term orientation, this alternative approach aims to include non-economic financial issues related to the social and environmental dimensions (Bui et al., 2020; Hsueh, 2014; Jamali et al., 2008; Wu & Pagell, 2011). Analyzing these dimensions through the lens of double materiality—now the prevailing approach according to recent European regulatory updates (see Sect. 2.4 and Chap. 3)—reveals increasing demands on corporate governance. Companies will need to address a growing range of environmental issues (e.g., deforestation, destruction of protected areas, climate change impacts) and social issues (e.g., hunger, poverty, migration, violent conflicts, child or forced labor). These issues may fall outside the company's internal reporting boundaries but have a direct impact on partners within the supply chain (Butt et al., 2019; Matos et al., 2020; Scherer & Voegtlin, 2020; Wang & Lotfi, 2024). Forms of collaboration between supply chain actors and stakeholder networks can help monitor these issues, especially considering internationally set targets such as the 17 Sustainable Development Goals (SDGs) of the UN 2030 agenda (Fiandrino et al., 2022; Gutierrez et al., 2022). While research has primarily focused on private corporations or listed entities, studies have outlined the risks that come with striving to achieve legitimacy in public procurement (Ghisetti, 2017), mainly due to procedural inconsistencies (Rogerson et al., 2024).

The effectiveness of governance in supply chains depends heavily on the context and nature of the relationships among actors. Pilbeam et al. (2012) presented a framework that connects governance instruments to network outcomes based on the surrounding context, thus shifting the attention from the dyadic relationship (i.e., two actors) to the meso-level one (i.e., more than two actors). These formal instruments—usually implemented in situations that are dynamic and unstable, such as organizational transitions—can produce outcomes such as improved coordination, control, viability, and performance. Conversely, informal instruments are typically used in settings where there are established relationships between the participants. In this regard, Bonatto et al. (2020) researched the contextual factors that influence the identification of relational governance instruments in supply chains and the estimated performance of these factors. In line with prior studies, they found that the choice of instrument is context-specific, while the performance obtained is mediated by trust and influenced by instruments characterized by flexibility, solidarity, and information sharing.

Given the need to account for the characteristics of the actors and the environment, it is difficult to uncover solutions that are a panacea for all problems. Thus, any delineated best practices may have limited external validity outside the context of their validation, as discussed by Martin and Patterson (2009), Sellitto (2018), or Soni and Kodali (2013). A further element of fragmentation is the lack of a system for collecting, validating, and processing data—and therefore information—across the various tiers of the supply chain (Barratt & Barratt, 2011). These linkages help address the lack of visibility of the 'structural holes' that inhibit information sharing across the supply chain decision-making process. Specifically, linkages promote the comparability of practices and performances, including best practices (Lotfi et al., 2013; Zhou & Benton Jr, 2007). This is particularly evident when all actors have a strong sustainability orientation (Cui et al., 2024; Khan et al., 2016b).

However, as discussed by several authors (Boström et al., 2015; Keating et al., 2008) and extensively covered in Sect. 3.5, there are multiple barriers to the implementation of sustainable supply chain governance. One of the most challenging aspects to overcome is the mounting complexity of supply chains. In this context, Nguyen and Zuidwijk (2025) recently conducted a literature review to confirm the relationship between achieving sustainable supply chain governance and the presence of supply chain complexity. Figure 4.3 reports their proposed relation.

This complexity, in particular, not only complicates the collection of sustainability information related to sustainability practices but also widens the gap between

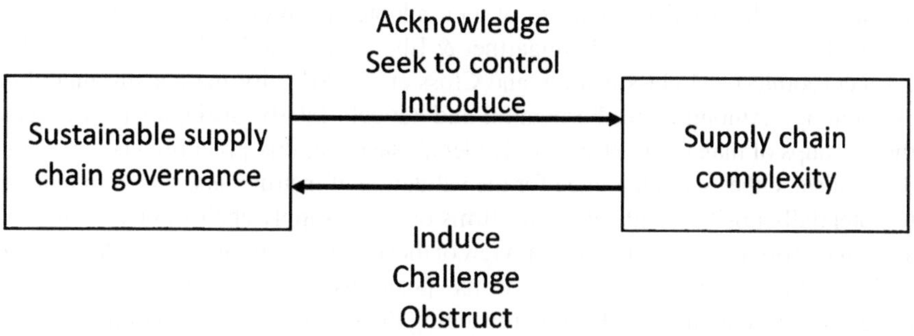

Fig. 4.3 Framework proposed by Nguyen and Zuidwijk (2025) to illustrate the mutually dependent relationship between sustainable supply chain governance and supply chain complexity. Source: Nguyen & Zuidwijk (2025, p. 557)

Europe and outside contexts, especially those untouched by recent directives (see Brockhaus et al., 2013, for a US-EU comparative study). Understanding these dynamics has significant implications for the accountability of supply chains and, by extension, the effectiveness of regulatory interventions (for the European context, see the work published by Vermeulen & Seuring, 2009).

In light of the above, analyses of globalized supply chains need to increase the awareness of risks and opportunities embedded in the design of long international supply chains. In this vein, authors such as Gereffi et al. (2005) or Amini and Bienstock (2014) have proposed theoretical frameworks to explain the historical evolution of this internalization trend. They assumed that modern supply chains can be governed by variables like the transaction complexity, the ability to codify the transaction, or suppliers' capabilities. What we can deduce is that the more power that focal actors have (i.e., the more the structure of the supply chain is hierarchical or relational), the more they can control the supply chain coordination (Gimenez & Sierra, 2013). Consequently, we can expect that a greater degree of sustainability orientation among supply chain actors will make data collection more effective—or, at least, that there will be higher pressure for the consistency and availability of data (Gereffi & Lee, 2016). Specific moments or disruptions, such as the 2008 financial crisis (Jüttner & Maklan, 2011) or the COVID-19 pandemic (Yu et al., 2021), have exposed how easily global supply chains can be disrupted by logistics and coordination difficulties, putting this globalization model in crisis (Durugbo & Al-Balushi, 2023; Ramakrishna & Wahab, 2023).

The board of directors (BoDs), CSR committee(s), and top management teams (or C-level in a broad sense) play key roles in mitigating these disruptions at both the firm (micro) and supply chain (meso) levels. Indeed, there appears to be a link between sustainability (outcome) performance and firms' propensity to disclose (for more details, see the next subsection). Within this debate, Benjamin et al. (2020) noted that the BoD's composition and gender diversity lead to an overall positive impact on sustainable supply chain responsibility, although the effect is lessened in the presence of CEO duality and the CSR committee. Scholars initially attempted to classify firms in terms of their adopted positioning within the sustainability domain (if any) (see Baumgartner & Ebner, 2010, or Walker & Jones, 2012, for a corporate-level classification and Closs et al., 2011, for market sustainability positioning). Among them, Formentini and Taticchi (2016) distinguished between three groups of management styles: 'leaders,' 'sustainability practitioners,' and 'traditionalists. Other scholars opted for the validation of instruments, such as a matrix, to potentially position and categorize firms or entire supply chains in terms of their sustainable orientation (for a recent review of those classifications, see Montag, 2023).

Given the competitive value of firms' positions on environmental and social issues, recent scholars have begun broadening their focus from BoD composition to the presence (or absence) of dedicated subcommittees, such as sustainability or CSR committees (Bolourian et al., 2023; García-Sánchez et al., 2019; Uyar et al., 2021). Authors such as Elmaghrabi (2021) have considered attributes of CSR committees that might enhance their impact, such as the number or percentage of independents, the presence of women (e.g., as chairs), the frequency of meetings, members' sustainability expertise, and the size of the committee. Naciti (2019) found that the presence of a CSR committee on a board significantly enhances a company's sustainability performance. Such committees are particularly effective in improving ESG outcomes when they have diverse members, especially with independent directors, and when there is active engagement from the company in sustainability issues. Gennari and Salvioni (2019) found that CSR committees on European boards are increasingly relevant to not only accounting but also corporate governance studies, given their role in fighting corruption, protecting stakeholders, and reducing exposure to responsibility failures. Relatedly, Sebastianelli and Tamimi (2020) explored the impact of CSR committees on corporate social performance. They found that companies with these dedicated committees often performed better in social responsibility areas, particularly when these committees were well integrated into the board's governance structure. Finally, Velte (2024) found that sustainable board governance significantly enhances sustainable supply chain reporting. This effect is particularly strong when boards include sustainability committees, achieve a higher number of female directors, and incorporate sustainability-related executive compensation. Collectively, these actions can improve environmental and social supply chain transparency in European firms.

Beyond the board, the top management team exerts considerable influence on sustainability performance. Aguilera et al. (2021) demonstrated that the top management team informs the sustainability orientation of corporate strategy. These individuals' commitment positively shapes the relationship between sustainability risk management and founding values, which is important to address both internal and external stakeholder pressures (Wijethilake & Lama, 2019). However, there is a need for more research that illuminates the CEO and CFO's roles in promoting ESG performance (Garcia-Blandon et al., 2019; Velte, 2020). Additionally, Iannuzzi et al. (2023) indicated that the likelihood of ESG controversies diminishes with increased overlap between the nomination committee and sustainability committee. Companies also reported fewer disputes when the director was younger or from a country other than the firm's headquarters. Furthermore, greater independence within the nomination committee correlates with a reduction in ESG- or sustainability-related conflicts. Notably, the controversial aspects of the social and environmental dimensions are often considered separately for their impact on the micro-level dimension (Pacelli et al., 2023).

Another relevant approach involves looking at the role of board members, the presence of a CSR committee, and the characteristics of the top management team (namely, C-suite executives) in unison rather than separately. Within this debate, scholars have also examined how the characteristics of board members and the sustainability committee—such as the roles of the CEO, CFO, female directors, and a higher percentage of independent directors—impact environmental and social controversies (Aouadi & Marsat, 2018; Cicchiello et al., 2023; DasGupta, 2022). In fact, Lewis et al. (2014) revealed that CEOs' (low) tenure and (financial) education increase their likelihood of disclosing the firm's environmental performance, which then affects supply chain actors' propensity to disclose. Meanwhile, Liu et al. (2023a) revealed that CEOs' foreign experience has a positive impact on their firms' ESG performance, which is further (positively) moderated by the CEO being female and more educated, as well as the firm being state-owned. Another important motivational attribute is how CEOs interpret and react to CSR as a competitive threat. According to Wang et al. (2023b), CEOs' awareness of how stakeholders have responded to peer firms' CSR practices affects their decisions to appoint CSR managers.

Another stream of scholarly discussion is the relation between ESG performance and the positive moderating role of CEO power (Velte, 2020). In line with this perspective, Guo et al. (2021) tested how educational similarity shapes the degree of CEOs' CSR disclosures. Since these top managers may be more inclined toward environmental disclosure, there is a risk of increased exposure to misreporting or misrepresenting significant environmental or social impacts. This, in turn, could negatively affect actual social and environmental (disclosure) performance. In this vein, Chatpibal et al. (2023) stressed the need to integrate ESG metrics into CFO performance evaluation criteria. However, more research is needed to better articulate the mechanisms, drivers, and barriers that link the CFO's presence on the CSR committee with greater accountability and transparency.

Meanwhile, further studies might find that the significance of the CEO's role is dependent on specific individual characteristics (Lewis et al., 2014), the presence of incentives (Ludwig & Sassen, 2022), or on the overall quality of the governance (Danish & Ulucak, 2020). Scholars have already uncovered a positive relationship between CEO and/or CFO positions and the quality of the firm's sustainability report (e.g., for a German sample, see Thun & Zülch, 2023), but they have yet to link those positions to ESG performance in the European or international context by broadening the area of investigation and the applied theoretical perspective (Orazalin, 2020; Uyar et al., 2021; Velte, 2020).

Given the growing importance of governance processes in transitioning toward more sustainable supply chain models, the next subsection of this chapter will explore the role of supply chain performance disclosure.

4.4 The Disclosure of Supply Chains' Sustainability Performance

The ongoing debate around the disclosure of corporate sustainability performance highlights the complexity and evolving nature of this topic. Scholars such as Cho et al. (2015a) have criticized the authenticity of sustainability reporting, describing it as an "organized hypocrisy" where companies project an image of responsibility without substantial actions, creating what they term "organizational façades." In another study, Cho et al. (2015b) revealed that these reports often do not reflect significant changes in corporate behavior, despite the growing emphasis on CSR disclosures, thus indicating a persistent gap between disclosure and actual practice. Aureli et al. (2020) further supported this skepticism by showing that the value relevance of ESG disclosures—particularly in companies listed on the Dow Jones Sustainability World Index—varies greatly and often fails to provide stakeholders with a clear picture of corporate sustainability performance. Building on the broader concerns about authenticity and value relevance in sustainability reporting, scholars have increasingly shifted their focus to the specific challenges and potential impacts of performance disclosure within complex supply chain networks.

Against this backdrop of "organizational façades" and uncertain value-relevance in standalone ESG reports (Aureli et al., 2020; Cho et al., 2015a, 2015b), this volume argues that moving from firm-level disclosure to multi-tier management, regulatory, and public transparency across global supply chains is essential for closing the disclosure–performance gap described above. In order to position this gap within the broader debate, Table 4.1 reports the four main types of transparency in value chains and networks, as proposed by Mol (2015). According to Mol (2015), management research originally framed "management transparency" as an internal, logistics-driven practice: Information moves mainly between upstream and downstream firms inside the chain, with neither citizen-consumers nor public authorities involved (first type). A second, "regulatory transparency" mode emerged when lawmakers required firms to pass specific data along the chain for inspectors and regulators, as in the EU's tracking-and-tracing rules (Mol, 2015, second type). The third

Table 4.1 Four types of transparency in value chains and networks

Ideal-type transparency	Who discloses information	Who receives information	Illustrative example
Management transparency	Upstream economic actors in the chain	Downstream economic actors in the chain	Total quality management
Regulatory transparency	Economic actors in the chain	Regulatory and inspection bodies	EU tracking-and-tracing system
Consumer transparency	Economic actors and certification bodies	Consumers and certification bodies	Eco-labels and product certifications
Public transparency	Economic actors and certification bodies	Public (citizen-consumers)	Carbon disclosure project

Source: Mol (2015, p. 155)

and fourth layers—often linked to eco-labels, certifications, and the rise of alternative networks—seek "consumer and public transparency," disclosing production and product details so that shoppers and certification bodies can judge sustainability claims (Mol, 2015). This volume situates itself squarely within that debate: It explores how new disclosure standards and due diligence obligations can deepen management- and regulator-led transparency while extending visibility to the public spheres, thus advancing the literature on supply chain disclosure and regulatory transparency.

Granted, achieving that transparency is easier said than done. As Bateman et al. (2017) argued, the disclosure and reporting of environmental sustainability in supply chains remain challenging due to the complexity of supply chain networks and the difficulties in measuring and communicating the relevant data. Kalkanci et al. (2016) noted that these disclosures are often manipulated to manage perceptions rather than to fully inform stakeholders. Qorri et al. (2018) developed a conceptual framework to measure sustainability performance in supply chains, highlighting the importance of integrating diverse metrics that capture the full spectrum of sustainability. Later, Qorri et al. (2022) introduced a practical method for assessing sustainability performance, even with incomplete information. Ultimately, they exposed the challenges that companies face in gathering comprehensive data across complex supply chains. Moreover, Diebel et al. (2024) revealed how suppliers respond to institutional pressures, particularly in global manufacturing networks. They found that voluntary environmental disclosures are often influenced by the complexity of institutional environments, which can either facilitate or hinder transparency. As effective disclosure remains one of the central challenges in advancing supply chain sustainability, Jia et al. (2024) emphasized the need for a clear roadmap to improve supply chain transparency. They argue that without greater clarity and consistency in reporting practices, efforts to enhance sustainability across global supply chains will continue to face significant obstacles. In order to partially solve this problem, Corazza et al. (2022) proposed that blockchain technology could bypass the barriers to sustainability disclosure in supply chains by providing a more transparent and traceable record of activities. Shen et al. (2025) showed that cross-border institutional forces shape suppliers' propensity to disclose carbon emissions and that this effect becomes stronger when the buying firm's carbon-management capabilities outpace those of its suppliers. In other words, a lead firm that is demonstrably better at managing carbon acts as a catalyst for upstream transparency (see also Amel-Zadeh & Tang, 2025, for the transition from voluntary to mandatory disclosure on the topic). Shen et al. (2025) further documented that the same globalization forces that nudge suppliers also raise the focal firm's own likelihood of disclosure, indicating a common mechanism at work across the chain. Taken together, the evidence supports a diffusion channel whereby international expectations for credible carbon information propagate through supply networks from lead firms to suppliers. Finally, Shen et al. (2025) find that greater disclosure by suppliers is associated with tighter, more collaborative buyer–supplier relationships. This perspective helps reconcile two strands of literature—global institutional pressure and supply chain governance—by showing that disclosure pressure does not stop at the firm boundary but "travels" along commercial ties. A key insight is the capability-transfer effect: superior carbon management at the lead firm is not just a signal to markets; it also becomes a governance lever that raises disclosure standards upstream via requirements, templates, and shared tools. For managers, this implies that investing in internal carbon systems (measurement protocols, data pipelines, assurance) yields multipliers beyond the firm, improving the quality and consistency of supplier data and, in turn, enabling more credible Scope 3 reporting. The finding that supplier transparency correlates with deeper collaboration suggests practical pathways (e.g., supplier training, joint data rooms, and harmonized KPIs) that both reduce transaction costs and build resilience.

Two caveats are worth noting. First, causality can be bidirectional: Firms that already maintain collaborative supply relations may be better positioned to elicit disclosure, so robust identification (e.g., instrumental variables or policy shocks) matters when generalizing. Second, industry and country heterogeneity likely condition the mechanism; heavily regulated or export-oriented sectors may experience stronger diffusion than domestically focused ones. Even with these precautions, the central message in this stream of literature is clear: Upgrading the lead firm's carbon management is a strategic prerequisite for credible, chain-wide transparency, especially as international stakeholders intensify demands for audited Scope 3 information.

This need for improved transparency and reporting is becoming even more pressing with the introduction of the European standards, namely the ESRS (for more details, see Sect. 3.3.3) and the CSRD (for more details, see Sect. 3.3.2). These regulations are expanding the scope of disclosure beyond corporate (firm-level) boundaries to include supply chain impacts, reflecting the growing recognition that sustainability cannot be adequately assessed without considering the broader network of a company's operations, including its suppliers and partners.

Moreover, the challenges of effectively disclosing sustainability performance extend beyond the traditional scope of corporate evaluation. As indicated by Papoutsi and Sodhi (2020), there is a question about whether sustainability reports truly reflect a company's actual performance or if they serve more as a tool for managing stakeholder perceptions. Deo et al. (2024) further argued that accounting practices in sustainability disclosures inform the dialogue with stakeholders, yet these practices often fall short in providing a complete and accurate representation of a company's environmental and social impacts.

Another critical factor to consider in the dynamics of sustainability orientation adoption is the relationship between firms and their supply chain partners. Caccialanza and Torelli (2024) revealed that industry associations provide platforms for shared learning and standard setting, which drive companies to align with industry-wide sustainability benchmarks. However, as noted by Caccialanza (2025), family firms may not respond to industry pressure as well. In fact, they tend to exhibit more selective disclosure practices at the governance level, where certain sustainability performances (particularly those that may not align with the firm's private or strategic interests) are less likely to be transparently reported. This disparity is why the literature underscores the importance of both micro-level (firm governance) and meso-level (associations) factors in shaping disclosure practices. Falcone et al. (2024) discussed how companies are more likely to adopt CSR practices when their customers and suppliers are closely connected. They suggest that connections in a closed network (i.e., local suppliers) amplify the positive influence of proactive, sustainability-oriented partners. As a result, firms have to directly engage in practices in order to reap the benefits (Falcone et al., 2024). Liang et al. (2023) showed that when downstream firms communicate an optimistic (or pessimistic) view of future demand in the forward-looking disclosures of their 10-K filings, upstream partners adjust their own resource commitments accordingly: the more upbeat the customer's tone, the more the supplier is willing to let costs "rise with sales" and, symmetrically, the less it cuts costs when sales fall (a pattern known as cost stickiness). Crucially, this response intensifies when the supplier produces highly specialized components for that customer, underscoring how tight, information-rich relationships reduce uncertainty and allow costs to be allocated where they create most of the value. The findings complement earlier evidence that managers modulate sticky costs in light of external signals (Anderson et al., 2003; Banker & Byzalov, 2014), but extend the argument to supply chain communication that is implicit (i.e., embedded in narrative disclosures) rather than explicit data-sharing protocols. Put differently, effective impact disclosure and cost allocation go hand in hand: When a lead firm is transparent about future prospects, suppliers gain the confidence to invest (or divest) capacity in line with genuine demand, improving overall supply chain efficiency. Conversely, opaque or 'boilerplate' forward-looking disclosures force suppliers to retain slack 'just in case,' raising system-wide costs and blunting incentives to mitigate environmental or social impacts.

Finally, the effects of new regulations like ESRS and CSRD extend beyond the EU, as the requirements placed on European firms also impact their global supply chains. This creates a ripple effect, in which non-EU firms involved in EU supply chains must also adapt to these stringent disclosure standards. The evolving regulatory landscape, particularly in the EU, will likely continue to shape the future of sustainability reporting, both within Europe and globally. While scholars have made significant progress in understanding and improving sustainability disclosures, numerous challenges and gaps remain. For instance, there is a need to explore the global implications of these regulations, particularly in terms of how non-EU firms will respond to the increased demand for transparency and accountability in their operations. Chapter 5 will discuss these arguments and their impact on data collection procedures and then arrange those insights into an actionable framework.

References

Abbasi, M. (2017). Towards socially sustainable supply chains–themes and challenges. *European Business Review, 29*(3), 261–303.

Adams, C. A. (2004). The ethical, social, and environmental reporting-performance portrayal gap. *Accounting, Auditing & Accountability Journal, 17*(5), 731–757.

Adams, C. A., & McNicholas, P. (2007). Making a difference: Sustainability reporting, accountability and organisational change. *Accounting, Auditing & Accountability Journal, 20*(3), 382–402.

Aguilera, R. V., Aragón-Correa, J. A., Marano, V., & Tashman, P. A. (2021). The corporate governance of environmental sustainability: A review and proposal for more integrated research. *Journal of Management, 47*(6), 1468–1497. https://doi.org/10.1177/0149206321991212

Aguilera, R. V., & Jackson, G. (2010). Comparative and international corporate governance. *The Academy of Management Annals, 4*(1), 485–556.

Ahi, P., & Searcy, C. (2015). An analysis of metrics used to measure performance in green and sustainable supply chains. *Journal of Cleaner Production, 86*, 360–377.

Ahmad, F., Bask, A., Laari, S., & Robinson, C. V. (2023). Business management perspectives on the circular economy: Present state and future directions. *Technological Forecasting and Social Change, 187*, 122182.

Ahmad, S., Wong, K. Y., & Rajoo, S. (2019). Sustainability indicators for manufacturing sectors: A literature survey and maturity analysis from the triple-bottom line perspective. *Journal of Manufacturing Technology Management, 30*(2), 312–334.

Allen, F., Carletti, E., & Marquez, R. S. (2009). *Stakeholder capitalism, corporate governance and firm value*. *EFA 2007 Ljubljana Meetings Paper, European Corporate Governance Institute (ECGI)—Finance Working Paper No. 190/2007, Wharton Financial Institutions Center Working Paper. https://ssrn.com/abstract=968141 or https://doi.org/10.2139/ssrn.968141

Al-Mawali, H. (2023). Proposing a strategy map based on sustainability balanced scorecard and DEMATEL for manufacturing companies. *Sustainability Accounting, Management and Policy Journal, 14*(3), 565–590.

Amaeshi, K. M., Osuji, O. K., & Nnodim, P. (2008). Corporate social responsibility in supply chains of global brands: A boundaryless responsibility? Clarifications, exceptions and implications. *Journal of Business Ethics, 81*(1), 223–234.

Amel-Zadeh, A., & Tang, Q. (2025). Managing the shift from voluntary to mandatory climate disclosure: The role of carbon accounting. *The British Accounting Review, 57*(2), 101594.

Amini, M., & Bienstock, C. C. (2014). Corporate sustainability: An integrative definition and framework to evaluate corporate practice and guide academic research. *Journal of Cleaner Production, 76*, 12–19.

Anderson, M. C., Banker, R. D., & Janakiraman, S. N. (2003). Are selling, general, and administrative costs "sticky"? *Journal of Accounting Research, 41*(1), 47–63. https://doi.org/10.1111/1475-679X.00095

Aouadi, A., & Marsat, S. (2018). Do ESG controversies matter for firm value? Evidence from international data. *Journal of Business Ethics, 151*, 1027–1047. https://doi.org/10.1007/s10551-016-3213-8

Appannan, J. S., Mohd Said, R., Ong, T. S., & Senik, R. (2023). Promoting sustainable development through strategies, environmental management accounting and environmental performance. *Business Strategy and the Environment, 32*(4), 1914–1930.

Aranda-Usón, A., Scarpellini, S., & Moneva, J. M. (2024). Dynamic capabilities for a "circular accounting" and material flows in a circular economy. *Resources, Conservation and Recycling, 209*, 107756.

Argiolas, A., Rawhouser, H., & Sydow, A. (2024). Social entrepreneurs concerned about impact drift. Evidence from contexts of persistent and pervasive need. *Journal of Business Venturing, 39*(1), 106342.

Arjaliès, D. L., Rodrigue, M., & Romi, A. M. (2023). "Come play with us!" a grassroots research agenda for accounting and the circular economy. *Accounting Forum, 47*(4), 497–524.
Asiaei, K., Bontis, N., Alizadeh, R., & Yaghoubi, M. (2022). Green intellectual capital and environmental management accounting: Natural resource orchestration in favor of environmental performance. *Business Strategy and the Environment, 31*(1), 76–93.
Aureli, S., Foschi, E., & Paletta, A. (2023). Management accounting for a circular economy: Current limits and avenue for a dialogic approach. *Accounting, Auditing & Accountability Journal, 38*(9), 291.
Aureli, S., Gigli, S., Medei, R., & Supino, E. (2020). The value relevance of environmental, social, and governance disclosure: Evidence from Dow Jones sustainability world index listed companies. *Corporate Social Responsibility and Environmental Management, 27*(1), 43–52.
Austin, J., Stevenson, H., & Wei-Skillern, J. (2012). Social and commercial entrepreneurship: same, different, or both?. *Revista de Administração, 47*(3), 370–384.
Bai, C., Zhu, Q., & Sarkis, J. (2024). Circular economy and circularity supplier selection: A fuzzy group decision approach. *International Journal of Production Research, 62*(7), 2307–2330.
Banker, R. D., & Byzalov, D. (2014). Asymmetric cost behavior. *Journal of Management Accounting Research, 26*(2), 43–79.
Barratt, M., & Barratt, R. (2011). Exploring internal and external supply chain linkages: Evidence from the field. *Journal of Operations Management, 29*(5), 514–528.
Bateman, A. H., Blanco, E. E., & Sheffi, Y. (2017). Disclosing and reporting environmental sustainability of supply chains. In *Sustainable supply chains: A research-based textbook on operations and strategy* (pp. 119–144). Springer International Publishing AG.
Battilana, J., & Lee, M. (2014). Advancing research on hybrid organizing-insights from the study of social enterprises. *Academy of Management Annals, 8*, 397–441. https://doi.org/10.1080/19416520.2014.893615
Baumgartner, R. J., & Ebner, D. (2010). Corporate sustainability strategies: sustainability profiles and maturity levels. *Sustainable Development, 18*(2), 76–89.
Bekier, J., & Parisi, C. (2023). Co-construction of performance indicators for a circular city and its relation to a local action net. *Accounting, Auditing & Accountability Journal*, (ahead-of-print). https://doi.org/10.1108/AAAJ-07-2022-5911
Benameur, K. B., Mostafa, M. M., Hassanein, A., Shariff, M. Z., & Al-Shattarat, W. (2024). Sustainability reporting scholarly research: A bibliometric review and a future research agenda. *Management Review Quarterly, 74*(2), 823–866.
Benjamin, S., Mansi, M., & Pandey, R. (2020). Board gender composition, board independence and sustainable supply chain responsibility. *Accounting & Finance, 60*(4), 3305–3339.
Blanco, C. C. (2021). Supply chain carbon Footprinting and climate change disclosures of global firms. *Production and Operations Management, 30*(9), 3143–3160. https://doi.org/10.1111/poms.13421
Bolourian, S., Alinaghian, L., & Angus, A. (2023). Exploring the role of board-level corporate social responsibility committees in corporate social responsibility performance: A configurational approach. *Journal of Business Research, 169*, 114280.
Bonatto, F., de Resende, L. M. M., & Pontes, J. (2020). Relational governance in supply chain: A systematic literature review. *Benchmarking: An International Journal, 27*(6), 1711–1741.
Bonatto, F., Resende, L. M. M. D., & Pontes, J. (2022). Supply chain governance: A conceptual model. *Journal of Business & Industrial Marketing, 37*(2), 309–325.
Borzaga, C., & Tortia, E. (2006). Worker motivations, job satisfaction, and loyalty in public and nonprofit social services. *Nonprofit and Voluntary Sector Quarterly, 35*(2), 225–248.
Boström, M., Jönsson, A. M., Lockie, S., Mol, A. P. J., & Oosterveer, P. (2015). Sustainable and responsible supply chain governance: Challenges and opportunities. *Journal of Cleaner Production, 107*, 1–7. https://doi.org/10.1016/j.jclepro.2014.11.050
Brockhaus, S., Kersten, W., & Knemeyer, A. M. (2013). Where do we go from here? Progressing sustainability implementation efforts across supply chains. *Journal of Business Logistics, 34*(2), 167–182.

Bruder, I. (2021). A social mission is not enough: Reflecting the normative foundations of social entrepreneurship. *Journal of Business Ethics, 174*(3), 487–505.

Bui, B., Houqe, M., & Zaman, M. (2020). Climate governance effects on carbon disclosure and performance. *The British Accounting Review, 52*(2), 1–16.

Butt, N., Lambrick, F., Menton, M., & Renwick, A. (2019). The supply chain of violence. *Nature Sustainability, 2*(8), 742–747. https://doi.org/10.1038/s41893-019-0349-4

Caccialanza, A. (2025). Sustainability performance disclosure and family businesses: A perspective article. *Journal of Family Business Management, 15*(1), 48–56. https://doi.org/10.1108/JFBM-10-2023-0250

Caccialanza, A., & Torelli, R. (2024). The role of trade associations in promoting corporate sustainability transition and reporting: A case study in the food supply chain. *Corporate Social Responsibility and Environmental Management, 31*(5), 4469–4486. https://doi.org/10.1002/csr.2807

Cagno, E., Negri, M., Neri, A., & Giambone, M. (2023). One framework to rule them all: An integrated, multi-level and scalable performance measurement framework of sustainability, circular economy and industrial symbiosis. *Sustainable Production and Consumption, 35*, 55–71.

Cantele, S., Moggi, S., & Campedelli, B. (2020). Spreading sustainability innovation through the co-evolution of sustainable business models and partnerships. *Sustainability, 12*(3), 1190.

Casati, M., Soregaroli, C., Frizzi, G. L., & Stranieri, S. (2024). Impacts of blockchain technology in agrifood: Exploring the interplay between transactions and firms' strategic resources. *Supply Chain Management: An International Journal, 29*(7), 51–70.

Castellano, N., De Luca, F., D'Onza, G., Maffei, M., & Melis, A. (2025). *Environmental, social, governance (ESG): Risk, performance, monitoring*. Springer Nature Switzerland AG. https://doi.org/10.1007/978-3-031-76618-3

Chatpibal, M., Chaiyasoonthorn, W., & Chaveesuk, S. (2023). *Driving financial results is not the only priority! An exploration of the future role of chief financial officer: A grounded theory approach*. Meditari Accountancy Research, (ahead-of-print). https://doi.org/10.1108/MEDAR-02-2023-1929

Chiu, T. T., Kim, J. B., & Wang, Z. (2019). Customers' risk factor disclosures and suppliers' investment efficiency. *Contemporary Accounting Research, 36*(2), 773–804.

Cho, C. H., Laine, M., Roberts, R. W., & Rodrigue, M. (2015a). Organized hypocrisy, organizational façades, and sustainability reporting. *Accounting, Organizations and Society, 40*, 78–94.

Cho, C. H., Michelon, G., Patten, D. M., & Roberts, R. W. (2015b). CSR disclosure: The more things change…? *Accounting, Auditing & Accountability Journal, 28*(1), 14–35.

Cho, C. H., & Patten, D. M. (2007). The role of environmental disclosures as tools of legitimacy: A research note. *Accounting, Organizations and Society, 32*(7–8), 639–647.

Chung, J. (2012). Measuring corporate environmental performance: The trade-offs of sustainability ratings. *Social and Environmental Accountability Journal, 32*(1), 39–39.

Ciambotti, G., Bacq, S., Haugh, H., Dorado, S., Doherty, B., Pedrini, M., & Markman, G. (2025). Transformational hybridity: Shape, shake, and shift up for societal grand challenges. *Journal of Management Studies, 62*(6), 2149–2168.

Cicchiello, A. F., Cotugno, M., & Foroni, C. (2023). Does competition affect ESG controversies? Evidence from the banking industry. *Finance Research Letters, 55*, 103972. https://doi.org/10.1016/j.frl.2023.103972

Claessens, S. (2006). Corporate governance and development. *The World Bank Research Observer, 21*(1), 91–122.

Closs, D. J., Speier, C., & Meacham, N. (2011). Sustainability to support end-to-end value chains: The role of supply chain management. *Journal of the Academy of Marketing Science, 39*, 101–116.

Conway Dato-on, M., & Kalakay, J. (2016). The winding road of social entrepreneurship definitions: A systematic literature review. *Social Enterprise Journal, 12*(2), 131–160.

Corazza, L., Zhang, J., Arachchilage, D. K., & Scagnelli, S. D. (2022). Blockchain and sustainability disclosure: A scenario-based application for supply chains. *Sustainability, 15*(1), 571.

Cornelissen, J. P., Akemu, O., Jonkman, J. G., & Werner, M. D. (2021). Building character: The formation of a hybrid organizational identity in a social enterprise. *Journal of Management Studies, 58*(5), 1294–1330.

Costa, E., & Andreaus, M. (2021). Social impact and performance measurement systems in an Italian social enterprise: A participatory action research project. *Journal of Public Budgeting, Accounting & Financial Management, 33*(3), 289–313.

Cui, X., Qi, B., & Hussain, M. J. (2024). Vendor sustainability performance and corporate customers' supplier selection. *Corporate Social Responsibility and Environmental Management, 31*(4), 2910–2928.

Curran, K., & Ozcan, P. (2024). Getting down to business: Governing the hybridization of UK charities. *Journal of Management Studies, 62*(6), 2169–2206.

Dal Maso, L., Gianfagna, L., Maglione, F., & Lattanzi, N. (2024). Going green: Environmental risk management, market value and performance. *Corporate Social Responsibility and Environmental Management, 31*(1), 122–132.

Danish, & Ulucak, R. (2020). The pathway toward pollution mitigation: Does institutional quality make a difference? *Business Strategy and the Environment, 29*(8), 3571–3583. https://doi.org/10.1002/bse.2597

DasGupta, R. (2022). Financial performance shortfall, ESG controversies, and ESG performance: Evidence from firms around the world. *Finance Research Letters, 46*, 102487. https://doi.org/10.1016/j.frl.2021.102487

De Nito, E., Caccialanza, A., Canonico, P., & Favari, E. (2024). Analyzing the role of social value in megaprojects: Toward a new performance framework. *Measuring Business Excellence, 28*(2), 243–254. https://doi.org/10.1108/MBE-09-2023-0138

Dees, J. G. (1998). *The meaning of social entrepreneurship*. Kauffman Center for Entrepreneurial Leadership.

Delmas, M., & Blass, V. D. (2010). Measuring corporate environmental performance: The trade-offs of sustainability ratings. *Business Strategy and the Environment, 19*(4), 245–260.

Deo, S., Pepper, M., De Zoysa, A., & Smark, C. (2024). Disclosures: Driving supply chain sustainability through accounting. In *Sustainable and resilient supply chain* (Vol. 12, pp. 49–64). Emerald Publishing Limited.

Diebel, W., Gualandris, J., & Klassen, R. D. (2024). How do suppliers respond to institutional complexity? Examining voluntary public environmental disclosure in a global manufacturing supply network. *Journal of Operations Management, 70*(2), 285–315.

Doherty, B., Haugh, H., & Lyon, F. (2014). Social enterprises as hybrid organizations: A review and research agenda. *International Journal of Management Reviews, 16*(4), 417–436. https://doi.org/10.1111/ijmr.12028

Dragomir, V. D. (2018). How do we measure corporate environmental performance? A critical review. *Journal of Cleaner Production, 196*, 1124–1157.

Durugbo, C. M., & Al-Balushi, Z. (2023). Supply chain management in times of crisis: A systematic review. *Management Review Quarterly, 73*(3), 1179–1235.

Eccles, R. G., Ioannou, I., & Serafeim, G. (2014). The impact of corporate sustainability on organizational processes and performance. *Management Science, 60*(11), 2835–2857.

Ecer, F., & Torkayesh, A. E. (2022). A stratified fuzzy decision-making approach for sustainable circular supplier selection. *IEEE Transactions on Engineering Management, 71*, 1130–1144.

Elmaghrabi, M. E. (2021). CSR committee attributes and CSR performance: UK evidence. *Corporate Governance: The International Journal of Business in Society, 21*(5), 892–919.

Falcone, E. C., Yan, T., & Fugate, B. S. (2024). Follow-suit or free-ride? A relational view of CSR diffusion in a supply chain with customer–supplier closure. *Journal of Operations Management, 70*(6), 979–1006. https://doi.org/10.1002/joom.1319

Fiandrino, S., Scarpa, F., & Torelli, R. (2022). Fostering social impact through corporate implementation of the SDGs: Transformative mechanisms towards interconnectedness and inclusiveness. *Journal of Business Ethics, 180*(4), 959–973.

Formentini, M., & Taticchi, P. (2016). Corporate sustainability approaches and governance mechanisms in sustainable supply chain management. *Journal of Cleaner Production, 112*, 1920–1933.

Galera, G., & Borzaga, C. (2009). Social enterprise: An international overview of its conceptual evolution and legal implementation. *Social Enterprise Journal, 5*(3), 210–228.

Garcia-Blandon, J., Argilés-Bosch, J. M., & Ravenda, D. (2019). Exploring the relationship between CEO characteristics and performance. *Journal of Business Economics and Management, 20*(6), 1064–1082.

García-Sánchez, I. M., Gómez-Miranda, M. E., David, F., & Rodríguez-Ariza, L. (2019). Board independence and GRI-IFC performance standards: The mediating effect of the CSR committee. *Journal of Cleaner Production, 225*, 554–562.

Gennari, F., & Salvioni, D. M. (2019). CSR committees on boards: The impact of the external country level factors. *Journal of Management and Governance, 23*(3), 759–785.

Gereffi, G., Humphrey, J., & Sturgeon, T. (2005). The governance of global value chains. *Review of International Political Economy, 12*(1), 78–104. https://doi.org/10.1080/09692290500049805

Gereffi, G., & Lee, J. (2016). Economic and social upgrading in global value chains and industrial clusters: Why governance matters. *Journal of Business Ethics, 133*(1), 25–38. https://doi.org/10.1007/s10551-014-2373-7

Ghisetti, C. (2017). Demand-pull and environmental innovations: Estimating the effects of innovative public procurement. *Technological Forecasting and Social Change, 125*, 178–187.

Giannakis, M., Dubey, R., Vlachos, I., & Ju, Y. (2020). Supplier sustainability performance evaluation using the analytic network process. *Journal of Cleaner Production, 247*, 119439.

Gimenez, C., & Sierra, V. (2013). Sustainable supply chains: Governance mechanisms to greening suppliers. *Journal of Business Ethics, 116*, 189–203.

Gimenez, C., & Tachizawa, E. M. (2012). Extending sustainability to suppliers: A systematic literature review. *Supply Chain Management: an International Journal, 17*(5), 531–543.

Gold, S., & Wieland, A. (Eds.). (2024). *The supply chain: A system in crisis*. Edward Elgar Publishing.

Govindan, K., Aditi, Kaul, A., Darbari, J. D., & Jha, P. C. (2023). Analysis of supplier evaluation and selection strategies for sustainable collaSboration: A combined approach of best–worst method and TOmada de Decisao Interativa Multicriterio. Business Strategy and the Environment, 32(7), 4426–4447.

Gu, J., Wang, J., Shi, X., & Xu, X. (2024). Investigating supply chain participants' circular economy action effects on firm financial performance from a stakeholder theory perspective. *Journal of Purchasing and Supply Management, 30*, 100903.

Gualandris, J., Klassen, R. D., Vachon, S., & Kalchschmidt, M. (2015). Sustainable evaluation and verification in supply chains: Aligning and leveraging accountability to stakeholders. *Journal of Operations Management, 38*, 1–13.

Guo, C., Luo, J., & Tian, G. (2025). Green wave goes up the stream: Green innovation among supply chain partners. *The British Accounting Review*, 101725.

Guo, J., Kim, S., Yu, Y., & Kim, J. Y. (2021). Does CFO accounting expertise matter to corporate social responsibility disclosure in 10-Ks? *Journal of Applied Accounting Research, 22*(5), 800–822. https://doi.org/10.1108/JAAR-07-2020-0137

Gutierrez, L., Montiel, I., Surroca, J. A., & Tribo, J. A. (2022). Rainbow wash or rainbow revolution? Dynamic stakeholder engagement for SDG-driven responsible innovation. *Journal of Business Ethics, 180*(4), 1113–1136.

Halari, A., & Baric, M. (2023). Exploring accountant's involvement in circular economy: Experiences and perspectives of practitioners. *Qualitative Research in Accounting & Management, 20*(4), 421–446. https://doi.org/10.1108/QRAM-03-2022-004

Hiller, J. S. (2013). The benefit corporation and corporate social responsibility. *Journal of business ethics, 118*(2), 287–301.

Hota, P. K. (2023). Tracing the intellectual evolution of social entrepreneurship research: Past advances, current trends, and future directions. *Journal of Business Ethics, 182*(3), 637–659.

Hsueh, C. F. (2014). Improving corporate social responsibility in a supply chain through a new revenue sharing contract. *International Journal of Production Economics, 151*, 214–222.

Hunt, S. D., & Morgan, R. M. (1996). The resource-advantage theory of competition: Dynamics, path dependencies, and evolutionary dimensions. *Journal of Marketing, 60*(4), 107–114.

Iannuzzi, A. P., Dell'Atti, S., D'Apolito, E., & Galletta, S. (2023). Nomination committee characteristics and exposure to environmental, social and governance (ESG) controversies: Evidence from European global systemically important banks. *Corporate Governance: The International Journal of Business in Society, 23*(6), 1314–1338. https://doi.org/10.1108/CG-03-2022-0119

Ilinitch, A. Y., Soderstrom, N. S., & Thomas, T. E. (1998). Measuring corporate environmental performance. *Journal of Accounting and Public Policy, 17*(4–5), 383–408.

Jamali, D., Safieddine, A. M., & Rabbath, M. (2008). Corporate governance and corporate social responsibility synergies and interrelationships. *Corporate Governance: An International Review, 16*(5), 443–459.

Jia, F., Li, K., Chen, L., Nazrul, A., & Yan, F. (2024). Supply chain transparency: A roadmap for future research. *Industrial Management & Data Systems, 124*(9), 2665–2688.

Jia, Y., Wang, T., Xiao, K., & Guo, C. (2020). How to reduce opportunism through contractual governance in the cross-cultural supply chain context: Evidence from Chinese exporters. *Industrial Marketing Management, 91*, 323–337.

Jüttner, U., & Maklan, S. (2011). Supply chain resilience in the global financial crisis: An empirical study. *Supply Chain Management: An International Journal, 16*(4), 246–259.

Kalaitzi, D., Matopoulos, A., Bourlakis, M., & Tate, W. (2018). Supply chain strategies in an era of natural resource scarcity. *International Journal of Operations & Production Management, 38*(3), 784–809.

Kalkanci, B., Ang, E., & Plambeck, E. L. (2016). Strategic disclosure of social and environmental impacts in a supply chain. In A. Atasu (Ed.), *Environmentally responsible supply chains* (Vol. 3, pp. 259–282). Springer. https://doi.org/10.1007/978-3-319-30094-8_13

Kamaludin, M. F., Xavier, J. A., & Amin, M. (2024). Social entrepreneurship and sustainability: A conceptual framework. *Journal of Social Entrepreneurship, 15*(1), 26–49.

Keating, B., Quazi, A., Kriz, A., & Coltman, T. (2008). In pursuit of a sustainable supply chain: Insights from Westpac banking corporation. *Supply Chain Management: An International Journal, 13*(3), 175–179. https://doi.org/10.1108/13598540810871217

Khan, M., Hussain, M., & Saber, H. M. (2016b). Information sharing in a sustainable supply chain. *International Journal of Production Economics, 181*, 208–214.

Khan, M., Serafeim, G., & Yoon, A. (2016a). Corporate sustainability: First evidence on materiality. *The Accounting Review, 91*(6), 1697–1724.

Kouhizadeh, M., Zhu, Q., & Sarkis, J. (2023). Circular economy performance measurements and blockchain technology: An examination of relationships. *The International Journal of Logistics Management, 34*(3), 720–743.

Kumar, A., Shrivastav, S., Adlakha, A., & Vishwakarma, N. K. (2022). Appropriation of sustainability priorities to gain strategic advantage in a supply chain. *International Journal of Productivity and Performance Management, 71*(1), 125–155.

Larrinaga, C., & Garcia-Torea, N. (2022). An ecological critique of accounting: The circular economy and COVID-19. *Critical Perspectives on Accounting, 82*, 102320. https://doi.org/10.1016/j.cpa.2021.102320

Lee, Y., Hu, J., & Lim, M. K. (2024). Revisiting circular economy indicators: A circular supply chain perspective. *Journal of Purchasing and Supply Management, 30*, 100941.

Letza, S., Sun, X., & Kirkbride, J. (2004). Shareholding versus stakeholding: A critical review of corporate governance. *Corporate Governance: An International Review, 12*(3), 242–262.

Lewis, B. W., Walls, J. L., & Dowell, G. W. (2014). Difference in degrees: CEO characteristics and firm environmental disclosure. *Strategic Management Journal, 35*, 712–722. https://doi.org/10.1002/smj.2127

Liang, P., Cavusoglu, H., & Hu, N. (2023). Customers' managerial expectations and suppliers' asymmetric cost management. *Production and Operations Management, 32*(6), 1975–1993. https://doi.org/10.1111/poms.13952

Liao, H., & Wen, Z. (2023). Capturing attitudinal characteristics of decision makers in multi-criterion sorting problems for performance evaluation of sustainable and circular suppliers. *Annals of Operations Research, 355*, 1–26.

Liu, Y., Fang, W., Feng, T., & Xi, M. (2023b). Environmental strategy, green supply chain integration and sustainable performance: Examining the synergistic effects. *Management Decision, 61*(9), 2603–2628.

Liu, Y., Zhang, F., & Zhang, H. (2023a). CEO foreign experience and corporate environmental, social, and governance (ESG) performance. *Business Strategy and the Environment, 33*(4), 3331–3355. https://doi.org/10.1002/bse.3647

Logue, D., Edwards, M., & McAllister, G. (2025). Hybridity in non-profits: Innovating for social value creation. *Journal of Management Studies, 61*(6), 2274–2301.

Lotfi, Z., Mukhtar, M., Sahran, S., & Zadeh, A. T. (2013). Information sharing in supply chain management. *Procedia Technology, 11*, 298–304.

Ludwig, P., & Sassen, R. (2022). Which internal corporate governance mechanisms drive corporate sustainability? *Journal of Environmental Management, 301*, 113780. https://doi.org/10.1016/j.jenvman.2021.113780

Lyon, T. P., Delmas, M. A., Maxwell, J. W., Bansal, P., Chiroleu-Assouline, M., Crifo, P., ... & Wijen, F. (2018). CSR needs CPR: Corporate sustainability and politics. *California Management Review, 60*(4), 5–24.

Martin, P. R., & Patterson, J. W. (2009). On measuring company performance within a supply chain. *International Journal of Production Research, 47*(9), 2449–2460.

Matos, S. V., Schleper, M. C., Gold, S., & Hall, J. K. (2020). The hidden side of sustainable operations and supply chain management: Unanticipated outcomes, trade-offs and tensions. *International Journal of Operations & Production Management, 40*(12), 1749–1770.

Moggi, S., & Dameri, R. P. (2021). Circular business model evolution: Stakeholder matters for a self-sufficient ecosystem. *Business Strategy and the Environment, 30*(6), 2830–2842.

Mohammed, A., Setchi, R., Filip, M., Harris, I., & Li, X. (2018). An integrated methodology for a sustainable two-stage supplier selection and order allocation problem. *Journal of Cleaner Production, 192*, 99–114.

Mol, A. P. (2015). Transparency and value chain sustainability. *Journal of Cleaner Production, 107*, 154–161.

Montag, L. (2023). Circular economy and supply chains: Definitions, conceptualizations, and research agenda of the circular supply chain framework. *Circular Economy and Sustainability, 3*(1), 35–75. https://doi.org/10.1007/s43615-022-00172-y

Naciti, V. (2019). Corporate governance and board of directors: The effect of a board composition on firm sustainability performance. *Journal of Cleaner Production, 237*, 117727.

Nguyen, L. T., & Zuidwijk, R. (2025). Sustainable supply chain governance: A literature review. *Business Ethics, the Environment & Responsibility, 34*(2), 541–564.

Nogueira, E., Gomes, S., & Lopes, J. M. (2025). Unveiling triple bottom line's influence on business performance. *Discover Sustainability, 6*(1), 43.

Orazalin, N. (2020). Do board sustainability committees contribute to corporate environmental and social performance? The mediating role of corporate social responsibility strategy. *Business Strategy and the Environment, 29*(1), 140–153. https://doi.org/10.1002/bse.2354

Pacelli, V., Pampurini, F., & Quaranta, A. G. (2023). Environmental, social and governance investing: Does rating matter? *Business Strategy and the Environment, 32*(1), 30–41. https://doi.org/10.1002/bse.3116

Papoutsi, A., & Sodhi, M. S. (2020). Does disclosure in sustainability reports indicate actual sustainability performance? *Journal of Cleaner Production, 260*, 121049.

Parent, M. M., & Deephouse, D. L. (2007). A case study of stakeholder identification and prioritization by managers. *Journal of Business Ethics, 75*, 1–23.

Peredo, A. M., & McLean, M. (2006). Social entrepreneurship: A critical review of the concept. *Journal of World Business, 41*(1), 56–65.

Pilbeam, C., Alvarez, G., & Wilson, H. (2012). The governance of supply networks: A systematic literature review. *Supply Chain Management: an International Journal, 17*(4), 358–376.

Pryshlakivsky, J., & Searcy, C. (2017). A heuristic model for establishing trade-offs in corporate sustainability performance measurement systems. *Journal of Business Ethics, 144*, 323–342.

Qorri, A., Gashi, S., & Kraslawski, A. (2022). A practical method to measure sustainability performance of supply chains with incomplete information. *Journal of Cleaner Production, 341*, 130707.

Qorri, A., Mujkić, Z., & Kraslawski, A. (2018). A conceptual framework for measuring sustainability performance of supply chains. *Journal of Cleaner Production, 189*, 570–584.

Ramakrishna, Y., & Wahab, S. N. (Eds.). (2023). *Handbook of research on designing sustainable supply chains to achieve a circular economy*. IGI Global.

Rasche, A. (2024). Bringing sustainability into the boardroom—structure, mindset, competencies. In S. Thomsen (Ed.), *Lærebog i Bestyrelsesledelse* (pp. 281–303). DJØF.

Rawhouser, H., Cummings, M., & Newbert, S. L. (2017). Social impact measurement: Current approaches and future directions for social entrepreneurship research. *Entrepreneurship Theory and Practice, 43*(1), 82–115. https://doi.org/10.1177/1042258717727718

Rinaldi, L., Unerman, J., & De Villiers, C. (2018). Evaluating the integrated reporting journey: Insights, gaps and agendas for future research. *Accounting, Auditing & Accountability Journal, 31*(5), 1294–1318.

Roberts, D., & Woods, C. (2005). Changing the world on a shoestring: The concept of social entrepreneurship. *University of Auckland business review, 7*(1).

Rogerson, M., Grosvold, J., & Alves, K. (2024). *Climate change and modern slavery in public procurement.* https://www.unseenuk.org/wp-content/uploads/2024/01/Procurement-modern-slavery-full-report-1.pdf

Saebi, T., Foss, N. J., & Linder, S. (2019). Social entrepreneurship research: Past achievements and future promises. *Journal of Management, 45*(1), 70–95. https://doi.org/10.1177/0149206318793196

Saha, A. K., & Khan, I. (2024). Sustainable prosperity: Unravelling the Nordic nexus of ESG, financial performance, and corporate governance. *European Business Review, 36*(6), 793–815.

Santos, F. M. (2012). A positive theory of social entrepreneurship. *Journal of Business Ethics, 111*(3), 335–351. https://doi.org/10.1007/s10551-012-1413-4

Scherer, A. G., & Voegtlin, C. (2020). Corporate governance for responsible innovation: Approaches to corporate governance and their implications for sustainable development. *Academy of Management Perspectives, 34*(2), 182–208.

Sebastianelli, R., & Tamimi, N. (2020). Antecedents of sustainable supply chain initiatives: Empirical evidence from the S&P 500. *Business and Society Review, 125*(1), 3–22.

Sellitto, M. A. (2018). Assessment of the effectiveness of green practices in the management of two supply chains. *Business Process Management Journal, 24*(1), 23–48.

Shen, H., Wu, H., & Zhou, Y. (2025). The influence of a globalized supply chain on corporate carbon management and disclosure. *The British Accounting Review, 54*, 101696.

Soni, G., & Kodali, R. (2013). A critical review of supply chain management frameworks: proposed framework. *Benchmarking: An International Journal, 20*(2), 263–298.

Thun, T. W., & Zülch, H. (2023). The effect of chief sustainability officers on sustainability reporting—A management perspective. *Business Strategy and the Environment, 32*(4), 2093–2110. https://doi.org/10.1002/bse.3238

Tsao, Y. C., Balo, H. T., & Lee, C. K. H. (2024). Resilient and sustainable semiconductor supply chain network design under trade credit and uncertainty of supply and demand. *International Journal of Production Economics, 274*, 109318.

Uyar, A., Kuzey, C., Kilic, M., & Karaman, A. S. (2021). Board structure, financial performance, corporate social responsibility performance, CSR committee, and CEO duality: Disentangling the connection in healthcare. *Corporate Social Responsibility and Environmental Management, 28*(6), 1730–1748.

Vegter, D., van Hillegersberg, J., & Olthaar, M. (2023). Performance measurement system for circular supply chain management. *Sustainable Production and Consumption, 36*, 171–183.

Velte, P. (2020). Does CEO power moderate the link between ESG performance and financial performance? A focus on the German two-tier system. *Management Research Review, 43*(5), 497–520. https://doi.org/10.1108/MRR-04-2019-0182

Velte, P. (2024). Sustainable board governance and sustainable supply chain reporting: European evidence. *Journal of Strategy and Management, 17*(2), 222–243.

Vermeulen, W. J., & Seuring, S. (2009). Sustainability through the market–the impacts of sustainable supply chain management: Introduction. *Sustainable Development, 17*(5).

Vinante, C., Sacco, P., Orzes, G., & Borgianni, Y. (2021). Circular economy metrics: Literature review and company-level classification framework. *Journal of Cleaner Production, 288*, 125090.

Walker, H., & Jones, N. (2012). Sustainable supply chain management across the UK private sector. *Supply Chain Management: An International Journal, 17*(1), 15–28.

Wang, N., Pan, H., Feng, Y., & Du, S. (2023a). How do ESG practices create value for businesses? Research review and prospects. *Sustainability Accounting, Management and Policy Journal, 15*(5), 1155–1177.

Wang, T., Fu, Y., Rui, O., & De Castro, J. (2023b). Catch up with the good and stay away from the bad: CEO decisions on the appointment of chief sustainability officers. *Journal of Management Studies, 61*(4), 1295–1326. https://doi.org/10.1111/joms.12926

Wang, Y., & Lotfi, M. (2024). How climate change and modern slavery interact in the supply chain: A conceptual model development through a systemic review. *Business Ethics, the Environment & Responsibility, 34*(4), 1516–1539.

White, L., Lockett, A., Currie, G., & Hayton, J. (2021). Hybrid context, management practices and organizational performance: A configurational approach. *Journal of Management Studies, 58*(3), 718–748.

Wijethilake, C., & Lama, T. (2019). Sustainability core values and sustainability risk management: Moderating effects of top management commitment and stakeholder pressure. *Business Strategy and the Environment, 28*(1), 143–154. https://doi.org/10.1002/bse.2245

Wu, Z., & Pagell, M. (2011). Balancing priorities: Decision-making in sustainable supply chain management. *Journal of Operations Management, 29*(6), 577–590.

Young, D. R., & Lecy, J. D. (2014). Defining the universe of social enterprise: Competing metaphors. *Voluntas: International Journal of Voluntary and Nonprofit Organizations, 25*(5), 1307–1332. https://doi.org/10.1007/s11266-013-9396-z

Yu, Z., Razzaq, A., Rehman, A., Shah, A., Jameel, K., & Mor, R. S. (2021). Disruption in global supply chain and socio-economic shocks: A lesson from COVID-19 for sustainable production and consumption. *Operations Management Research, 15*, 1–16.

Zhou, H., & Benton, W. C., Jr. (2007). Supply chain practice and information sharing. *Journal of Operations Management, 25*(6), 1348–1365.

Chapter 5
The Theoretical Framework and Research Agenda on Sustainability and Accountability of Supply Chains

As anticipated in the introduction section, the logical framework presented herein (see Fig. 5.1) reflects the general structure of the book. This framework will help facilitate a clearer visualization of both the theoretical and practical implications arising from the implementation of the new regulatory framework in the EU and non-EU contexts. Similar to the structure of this volume, Fig. 5.1 is divided into three main thematic areas: the sustainability of value chains, the accountability of value chains, and the institutional level of intervention. Each area will receive a dedicated subsection to highlight the topic's components and related research agenda. The following discussion will pinpoint the most promising research avenues, including the interrelationships among the components within each area. Consequently, for each area, a series of research questions is proposed that can be further developed in future studies.

5.1 Sustainability of the Value Chain

The first area under discussion—pertaining to the "sustainability of the value chain"—corresponds to the book's opening subsection about the meso level (i.e., the supply chain) dimension of analysis. In particular, this section revisits the orientation toward sustainability within the broader supply chain management (i.e., the SSCM) literature (Koberg & Longoni, 2019).

The sustainability of the supply chain is not directly tied to the accounting literature or the accountability of production processes; instead, it is anchored by certain drivers and barriers. This aspect has dual significance: within the SSCM scholarship, it corresponds to the research stream on supply chain sustainability, but it also intersects with the institutional dimension by incorporating the drivers and barriers of sustainability reporting in accordance with the new ESRS standards. In line with

A. Caccialanza, *Sustainability and Accountability of Supply Chains*, SIDREA Series in Accounting and Business Administration,
https://doi.org/10.1007/978-3-032-16879-5_5

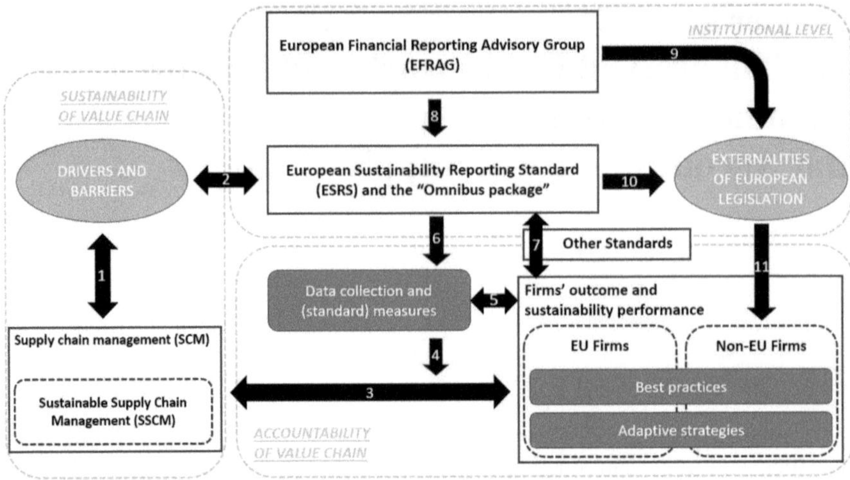

Fig. 5.1. Framework of the effects on the sustainable supply chain management (SSCM) debate of the new ESRS draft proposed by EFRAG. *Source*: Author's elaboration

the first relationship (bidirectional black arrow number 1 in the framework), we can derive the following research questions:

- How have the drivers and barriers to sustainability reporting evolved over time in relation to the increasing adoption of sustainability practices within the supply chain?
- What have been the main drivers of sustainability reporting in relation to the evolution of sustainability practices within the supply chain?
- What have been the main barriers to sustainability reporting in relation to the evolution of sustainability practices within the supply chain?

Similarly, following Wolf (2014), Govindan et al. (2021), and Guidi et al. (2024), we believe that this conversation can be expanded by addressing evolution in terms of the stakeholder(s) involved:

- How have the perceptions of different stakeholder categories driven or inhibited sustainability reporting practices over time in relation to the increasing adoption of sustainability practices within the supply chain?
- What have been the main drivers of sustainability reporting in relation to the evolution of sustainability practices within the supply chain, according to the perceptions of different stakeholder categories?
- What have been the main barriers to sustainability reporting in relation to the evolution of sustainability practices within the supply chain, according to the perceptions of different stakeholder categories?

Taking this further, we can draw from the research of El-Garaihy et al. (2024) and Grabs et al. (2024) to pose questions about the relationship between pressures and stakeholders in supply chains:

- How has stakeholders' awareness of the need to promote supply chain sustainability evolved over time?
- In what ways have each category of stakeholder adopted different drivers and levers to promote supply chain sustainability?
- What are the most recurring attributes or characteristics, depending on whether the stakeholders are internal or external?

This perspective considers both the managerial practices themselves and the actors who can implement, facilitate, or hinder the adoption of these practices. Thus, the chapter delineates the implications in terms of not only the scope of corporate action or influence in performance reporting but also the impact at the supply chain level.

The second relationship depicted in Fig. 5.1 (i.e., bidirectional black arrow number 2 in the framework) captures the relationship between the two dimensions involved and their components that are directly involved: drivers and barriers of sustainability reporting and the new ESRS.

Regarding the first type of relationship, a growing body of literature has explored how managerial practices are closely linked to regulatory and institutional influences (Daniel et al., 2012; Delmas & Toffel, 2004). The SSCM literature has settled into two general perspectives on this relationship: First, research has focused on understanding the managerial practices that drive this orientation within firms (Ahi & Searcy, 2013; Rajeev et al., 2017); second, scholars have sought to identify the institutional factors that most effectively encourage the adoption of such practices (Beske & Seuring, 2014; Pagell & Shevchenko, 2014; Trianni et al., 2017). Section 1.3 already outlined this debate, but there is ample room for further inquiry given the numerous ongoing regulatory initiatives. Building on the regulatory developments presented in Chapter. 3.1 of this volume, we can formulate several research questions. First, given that structural barriers (e.g., infrastructure limitations, compliance costs, and supply chain inertia) hinder equitable value distribution (Abbasi & Nilsson, 2012; Govindan & Bouzon, 2018):

- What are the (new) barriers preventing an equitable distribution of value within supply chains under increasing regulatory pressures for sustainability reporting according to the CSRD and ESRS?

Second, firms transitioning to sustainable supply chains must balance regulatory compliance, operational adjustments, and financial risks—an effort that impacts how value is redistributed (Khalid et al., 2015; Troisi et al., 2024). Accordingly, one can ask:

- What is the impact of sustainability reporting requirements on the decision-making processes of firms transitioning toward sustainable supply chain models, particularly in relation to value redistribution?

Next, we shift our attention to the accountability of supply chains and the role of institutional pressures, focusing on the interconnections depicted by the black arrows in Fig. 5.1.

5.2 Accountability of the Value Chain

The second chapter of this book shifted the focus from defining the supply chain dimension to evaluating its accountability to fulfill stakeholders' information pressures. Traditionally, accountability has been associated with corporate performance; however, the discussion here extends to evaluating the transparency of supply chain outcomes. For this reason, the second thematic area has been titled "Accountability of the Value Chain." This dimension entails assessing the extent to which value is transmitted along supply chains that contribute to the production of goods and/or services in the most transparent way. As pointed out by Gold and Heikkurinen (2018), firms can be opaque about the disclosure of their results or the accountability of their production systems, leading to a "transparency fallacy." In connection with the first thematic area ("Sustainability of the Value Chain"), this second area underscores the interrelation between supply chain operations and the mechanisms through which value is created and transferred within them.

The key relationship to understand is how specific aspects of a firm's internal operations can influence—or be influenced by—sustainability practices promoted at the supply chain level (Caccialanza & Torelli, 2024; Galli et al., 2024). These dynamics can be framed in two ways: The first concerns best practices within the firm; the second pertains to the adaptive strategies that firms adopt in response to one or more external stimuli or incentives (Han et al., 2024). In the first case, firms are seen as the key actors and initiators of change, either directly by engaging their suppliers and clients in sustainability initiatives or indirectly by promoting their own sustainability processes and hoping to trigger a shift in orientation. In the second case, an external stimulus drives a firm's strategic choices; firms may react passively, voluntarily, or proactively based on their own inclinations and resources.

This situation is further complicated by the EU's mandatory approach to supply chain regulation (Leite et al., 2024), since entrepreneurs and strategies are influenced by it (Busch et al., 2024; Moggi et al., 2022). To achieve this, the research agenda needs to go beyond the micro level and focus on assessing performance at the meso level. This agenda aims to structure future discussions on how accountability mechanisms operate across different levels of analysis and how they interact with the evolving regulatory and sustainability reporting frameworks. In this sense, the framework proposed herein considers both the dialogic mechanisms between the two major dimensions (i.e., the black arrow number 3) and the intervening role of performance metrics in the specific (moderating) effect of arrow number 4. This relationship, highlighted by arrows 3 and 4, underscores the importance of assigning value to the performance achieved at both the micro- and meso-levels. In parallel, Sect. 4.2 discusses the evolution of sustainability metrics from a micro level, or firm perspective to a broader supply chain meso level. This shift highlights the increasing need for standardized performance evaluation mechanisms to ensure meaningful sustainability integration, as indicated by arrow number 4. However, the absence of universally adopted standards continues to challenge the comparability and reliability of sustainability performance assessments (Adams, 2004; Adams &

McNicholas, 2007). On that basis (and in line with the black arrow number 4 in Fig. 5.1), we can formulate the following research questions:

- How do the evolving sustainability performance metrics in supply chain management impact firms' sustainability outcomes and strategic decision-making?
- How does the adoption of standards such as ESRS inhibit or facilitate the comparability of metrics that have already been proposed in the literature or adopted in managerial practice?
- Are there specific areas of sustainability, or subtopics therein, that are more suited to the comparability of metrics and performances?
- What factors constrain the transition from measuring performance at the micro level to the meso level?
- Are there enabling factors for this transition that have already been discussed in the literature? If so, which ones?

By doing so, this agenda provides a foundation for further empirical and theoretical investigations into the intersection of corporate practices implemented in a specific supply chain, as well as regulatory compliance in sustainability disclosure using targeted standards and metrics. The bidirectional black arrow number 5 in Fig. 5.1 captures this relationship. The primary goal is to bridge the gap between existing theoretical and empirical studies on these relationships and evolving metrics. In particular, arrow number 5 focuses on how sustainability reporting is increasingly shaping both disclosure practices and the communication of corporate performance in this new regulatory environment.

Among the trends driving both comparability and the development of new metrics, two specific topics have received deeper examination: the unique context of circular supply chains and the growing spread of new technologies. The adoption of circular economy (CE) principles has led to methodological innovations in sustainability reporting, with new performance measurement frameworks emphasizing closed-loop supply chain operations (Bai et al., 2024; Vegter et al., 2023). However, integrating CE-based metrics into firm-level decision-making is still a developing area that can support several potential inquiries:

- How does the integration of CE performance measurement systems influence firms' sustainability performance within supply chains?
- How do performance measurement standards enable performance comparability across firms?
- What are the best-practice examples in this regard from both theoretical and practical perspectives?
- What strategies are currently in place for data collection between EU and non-EU companies to ensure data integrity?
- Are there adaptive strategies that companies have enacted to improve the transparency and accountability of these measures?

With regard to data collection, there are numerous factors that can positively enhance firms' ability to collect, store, and process data. New technologies like blockchain and AI offer potential solutions for improving transparency in sustainability reporting, but their implementation remains fragmented (Beske-Janssen et al., 2023; Kazancoglu et al., 2023). Thus, researchers might ask:

- In what ways does the implementation of digital technologies (e.g., blockchain, AI, and interoperable platforms) influence the effectiveness of sustainability reporting, especially in terms of ensuring fair value distribution within supply chains?

Another key issue concerns institutional pressure, specifically how institutionalized habits have already been honed among practitioners. Specifically, we should focus on the efforts undertaken by other international frameworks and sources of institutional pressure, which illustrate a set of established habits between (institutional) proponents and (management and practitioners) adopters. Even before the introduction of the CSRD and ESRS, there were efforts to create uniform performance standards (such as the GRI), at least in the European context. The two-way relationship emphasized by arrow 7 of the proposed framework reflects this pressure to adopt international standards or performance measurement protocols, which has a direct impact on firms' perceptions of certain targets and performance metrics. That said, these standard setters also exert an indirect influence, as seen in the dialogue between existing standards and previously adopted protocols or agendas (e.g., the UN 2030 Agenda or the Integrated Reporting initiatives).

Building on these institutional pressures and standard-setting initiatives, the evolution of integrated reporting has progressively embraced deeper sustainability considerations within corporate strategies and supply chains. In fact, the evolution of integrated reporting has been marked by increasing recognition of the need to integrate sustainability deeply into corporate strategies and supply chains. Future proposals for integrated reporting suggest a greater focus on practical implementations and verifiable impacts, particularly in how organizations manage their supply chains in alignment with broader sustainability goals, at least for specific contexts (Caccialanza & Marinoni, 2023). On the one hand, the literature calls for a shift from theoretical frameworks to actionable strategies that embed sustainability in core business practices (Brown & Dillard, 2014; Lai & Stacchezzini, 2021). This might be linked to the conceptualization already proposed by Adams (2017), as retraced in Fig. 5.2.

On the other hand, organizations like EFRAG and ISSB aim to harmonize sustainability reporting, but scholars continue to debate their ability to reduce reporting inconsistencies and promote fair value distribution (Camilleri, 2020; Pickering et al., 2020). This misalignment is, in fact, inevitable due to the lack of overlap (the interoperability) between the various standards. This topic is represented in the framework by the black arrow number 7, which specifically focuses on the hypothesis of examining the ESRS alongside other existing standards while also considering performance in a broader sense or beyond the data points proposed by this predominantly European framework. As a result, firms experience friction (at least potentially) from two directions: the initial difficulty of implementing the new standard and the effort to maintain effective and efficient applications of prior standards.

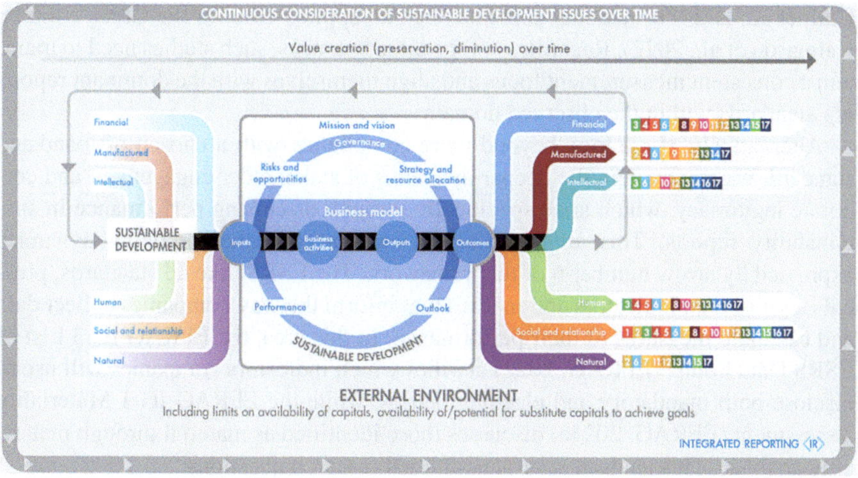

Fig. 5.2 Aligning SDGs with the value-creation process. *Source*: Adams—Integrated Reporting (2017, p. 14)

Thus, there are several relevant questions about how to manage the transition between standards:

- What role do international standard-setting bodies (EFRAG, ISSB) play in harmonizing sustainability reporting frameworks to facilitate a fairer distribution of value across global supply chains?
- How do sustainability reporting frameworks enable interoperability between different reporting systems?
- What validated tools and mechanisms have been implemented to facilitate the transition from an established reporting standard (e.g., Integrated Reporting) to a harmonized framework tailored to a specific regulatory context (e.g., ESRS for the EU)?

To advance this debate, scholars could pursue a literature review built around disentangling the existing dynamics that lead to interoperability between reporting systems. Future studies could critically assess their advantages and limitations, as well as the key drivers and barriers that influence their reporting systems' application across different contexts and mandatory reporting obligations. Fortunately, the transition from GRI to ESRS already provides a useful case study on how operators move to comply with EU regulations (see the operational guide Global Reporting Initiative, 2023). However, there is currently no comprehensive mapping of how the ESRS aligns with the GRI framework. Such a mapping could ensure comparability between the EU standard and one of the most widely recognized reporting frameworks. There is also limited evidence about the comparability of ESRS and GRI materiality assessment processes (Bayne, 2022; Chief Sustainability Officers Network, 2024; Turner & Weirich, 2023) as presented in 2.4. This analytical approach could be developed around specific sectors (e.g., the agri-food industry;

Galli et al., 2024) or national contexts (e.g., an applied case for the Greek context; Kalogirou et al., 2022). Regardless of the circumstances, such studies need to maintain a consistent measurement focus and align themselves with the dominant reporting standards within the observed domain.

Alternatively, if we look beyond mere compliance with a chosen or mandated standard, we can explore the deeper dynamics of stakeholder engagement and corporate legitimacy, which arise specifically through disclosing performance in sustainability reports. This area recalls, at least in part, the top-down dynamics expressed by arrow number 6 of the framework. The ESRS, like all standards, presents a set of data indicators and metrics that inform the way companies collect data and calculate measures of their performance. In this area, the EFRAG IG 3 List of ESRS Data Points (EFRAG, 2022) clarifies which indicators companies will use to disclose both mandatory and elective metrics, while the EFRAG IG 1 Materiality Assessment (EFRAG, 2024a) discusses those identified as material through materiality analysis. This leads to several additional research questions:

- How will the quality of reporting change with the adoption of new ESRS' data points?
- How should practitioners manage the adoption of new standards that overlap with prior ones (i.e., GRI)?
- Will the standards allow comparability at the supply chain level, or will they be limited to the firm level?
- Will the increased comparability have actual consequences with respect to supply chain accountability?
- Is greater accountability also reflected in a greater level of business legitimacy? If so, can that level be differentiated based on those who were already reporting according to other standards and those who only began to report in response to the ESRS?

Amidst the spread of sustainability reporting practices, compliance has become a strategic tool for accountability and stakeholder engagement (Appannan et al., 2023; Cho & Patten, 2007). Firms are increasingly integrating sustainability disclosures into their core business functions, but research has yet to clarify the effectiveness of these disclosures in improving supply chain sustainability performance. There is room for inquiries such as

- To what extent does sustainability reporting in supply chains enhance firms' overall performance and accountability beyond compliance obligations?

The next section addresses these compliance issues in more detail as part of a discussion of the institutional factors.

5.3 The Intervening Role of the Institutional Level: EFRAG and ESRS Standards

The third and final dimension analyzed in this book is the "institutional level," namely, the factors and entities that intervene in the relationship between the company's performance and the supply chain's sustainability. It is impossible for Fig. 5.1 to summarize all the regulations and standards that govern sustainability performance reporting in the European context. We direct interested readers to Sect. 3.1 for a full review; here, we deliberately focus on the dynamics that will intervene with the implementation of CSRD through the application of ESRS.

As illustrated in Sect. 3.2, the literature has created different approaches based on the area of inquiry rather than a unique reporting standard framework. Historically, scholars have struggled to design a uniform approach that can resolve all political and regulatory concerns; meanwhile, firms have devised a wide array of operational and reporting practices. Discrepancies in national regulations create inconsistencies that impact supply chain transparency (Das, 2017; Pickering et al., 2020)—and, by extension, deficiencies in measurement, comparability, and accountability. This leads to the following potential research question:

- To what extent do variations in national and international regulatory frameworks impact the uniformity and comparability of sustainability reporting practices across supply chains?

Of course, it is already difficult to perform large-scale analyses in multiple sectors, much less in multiple countries. Thus, there are substantial challenges in doing a priori estimates on the effects of European legislation on non-EU contexts that are not subject to these mandatory constraints. Scholars such as Garcia-Torres et al. (2024) have attempted to reconstruct these impacts at the level of a single sector, such as fashion apparel, in a focal firm's supply chain. However, comparative analyses will only be possible with the cultivation of a greater propensity for data disclosure, as well as the dissemination of comparable standards that can address the fragmentation seen up to now.

In this context, the role that the new standards promoted by EFRAG have in informing the choices of companies with respect to the possibility of implementing sustainability reporting metrics for mandatory disclosure purposes is evident. The connection between the ESRS standard and the process of data collection and standardization of the measures is expressed by the unilateral black arrow number 6, an expression of the unambiguousness of the interpretation of those numerical metrics.

There have been several EFRAG reports and grey literature publications on firms' preparedness with regard to implementing ESRS. For instance, PwC's recent report (2024) analyzed the state of readiness in hundreds of companies worldwide by means of surveying their top managers' perceptions of the implementation. Figure 5.3 reveals a fairly fragmented view of the perceived scope of preparedness. Specifically, managers seemed to express greater confidence in those topics that were more familiar and for which regulatory interventions were more established,

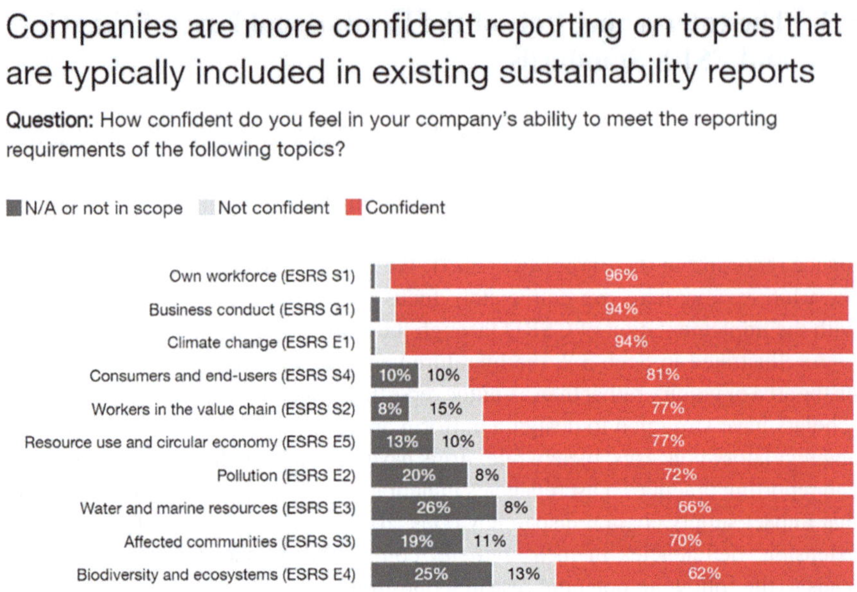

Fig. 5.3 Companies are more confident reporting on topics that are typically included in existing sustainability reports (n = 547). *Source*: PwC Global CSRD Survey 2024

such as "own workforce" or "business conduct." On the contrary, they seemed to express more concern for all those topics where there was a lower degree of control over the variables—and therefore less expertise in reporting or evaluating the impact on the topic. Recent studies (see Dutta & Dutta, 2024, for the Finnish context) have already highlighted the correlation between environmental performance and firms' propensity to disclose biodiversity impacts.

According to PwC's (2024) analysis, the areas demanding the most attention—due to lower corporate confidence in reporting—are precisely those linked to impacts outside the organization. While internal workforce dynamics are perceived as being more easily controlled, the greater the distance from direct oversight, the more challenging it becomes to estimate and potentially mitigate these impacts. This kind of social and environmental impact assessment is closely connected to, and widely debated in, the literature on socio-environmental accounting (Brown et al., 2021; Moggi, 2019). However, these interconnections must also be considered while evaluating the impacts within the production chain (Bales & Sovacool, 2021). Confidence in reporting certain meso level and/or environmental/social topics can shift the perception of related issues, particularly when they are tied to an already addressed business concern—such as the availability of standards in a specific sector or country (Clarkson et al., 2008; Luo et al., 2013; Scagnelli et al., 2013). Based on these considerations, we can formulate the following proposals for further analysis:

- How are firms' perceptions of internal expertise changing in response to ESRS implementation?
- Are these perceptions different among companies that previously applied an international reporting standard?
- Did previous experience with a standard cause the application of ESRS to have a substitution effect, or did firms attempt an integrated application of the two standards?
- Have firms' perceptions about applying ESRS faced different dynamics based on certain contextual factors? Specifically, did companies in industries with high environmental and/or social impacts implement timely training to offset the potential legitimacy gap in their activities?
- Has the quality of institutions impacted the implementation of ESRS at the firm level?

There may be additional elements that characterize the enterprise itself rather than the industry or other macroeconomic factors. These are expressed by arrow 6, which captures the connection between organizational and institutional levels in terms of performance accountability. The literature has already found that a firm's size influences its level of preparedness for regulatory change, as does its orientation toward sustainability or circular business models. Additional considerations and proposals for this research agenda can also be evaluated by considering EFRAG's specific guidelines for sustainability performance reporting, designed to promote disclosure among SMEs. Published in December 2024, these Voluntary Reporting Standard for SMEs (VSMEs) guidelines include supplementary information on their application and significantly fewer indicators, reflecting the size thresholds of the undertaking involved (EFRAG, 2024b). Arkoh et al. (2024) suggested that scholars need to pay more attention to the analysis of non-listed and smaller firms' propensity to publish sustainability reports. On this basis, some topics for further development include:

- Are SMEs significantly more challenged in reporting sustainability performance?
- What effect has the publication of guidelines, such as VSMEs, had on smaller firms' propensity to report sustainability performance?
- How has the VSME guideline publication shaped firms' perceptions about their own ability to report their sustainability performance?
- Are there additional impediments that could be resolved with further clarification from the EFRAG? If so, what are they, and what are the proposed remedies?

Shifting attention from application to conceptualization, further research could focus on aspects related to the design of standards, especially from the perspective of the standard setter (here, EFRAG). For instance, one could make the following inquiries, in line with the top-down arrow number 8 in Fig. 5.1:

- What were the dynamics that led to the approval of the ESRS in its current version?

- What logic took priority when designing the standards? The performance logic of optimizing the technical facets of production chains or the political logic of spotlighting the issue without burdening businesses too much, too soon?

Since the EU is part of a highly interdependent context with respect to other major economies and trading blocs, there is an important issue regarding externalities. Arrow number 9 in Fig. 5.1 reflects the expectation that European legislation has indirect effects on non-European entities. This situation raises several questions:

- What (additional) benefits will result from expanding the level of disclosure to the whole supply chain?
- Will the regulations related to CSRD implementation, through the application of ESRS, really allow a benefit for EU companies, or will they constitute an annual bureaucratic burden?

Another major issue is the effects imposed on firms that are not headquartered in one of the EU member countries but nevertheless have a considerable amount of business within the European economic area (as outlined in Sect. 3.3 of this book). In this sense, the new standards have international implications. Thus, the following questions may be useful to consider:

- To what extent were the competitive implications for non-EU firms evaluated during the approval of these norms?
- Will firms respond to these disclosure requirements by following the logic of compliance, or will they actually innovate their business models?
- As a result, does the modular approach—incorporating size-related obligations and phased deadlines for progressively smaller thresholds—ensure that the regulations' anticipated impact remains aligned with the principle of proportionality?

Regarding firm size, scholars have already found that some firms respond more proactively than others in terms of transitioning toward sustainability or even adopting circular business models (Ahmadov et al., 2023; Neri et al., 2023). That said, it would be interesting to analyze whether companies that fall outside of the European regulations (at least directly) are encouraged to adopt performance reporting schemes that conform to European ones. Arrow number 10 in Fig. 5.1 captures the relationship between the externalities of European regulation and the performance measurements of non-EU companies. Accordingly, we consider the following group of research questions:

- Which companies would have an interest in applying European reporting standards even though they are not directly subject to them?
- What characteristics do these firms have?
- Are there commonalities with the similar behavior of other European firms adopting similar schemes, such as simplified VSMEs?

Finally, as represented by arrow number 11 of the proposed framework, scholars could pursue theoretical works or case study analyses of the direct externality effects of European regulations on non-EU businesses. On the one hand, these studies could explore the best practices of non-EU firms that have adopted performance measurement systems, especially in view of their interoperability with the European system. On the other hand, studies could examine the adaptive logics that businesses have applied to their performance measurement strategies. Such an investigation would illuminate the strategic differences between EU and non-EU firms, as well as expose how performance is expressed at both the micro and meso levels.

Obviously, it is already difficult to achieve a clear, transparent, and comparable measurement at both the internal and cross-sectoral levels within the EU; it becomes even more challenging to raise expectations for further comparability at the meso level. This applies not only to a group of businesses within the same sector but also to those with greater internal heterogeneity, both in terms of sectoral and size-related differences. Although the level of accountability for performance at both the micro and meso levels is not yet at the desired or optimal stage, regulators and practitioners are working to balance the demands for accountability with the limited resources available. By proposing such complex or idealistic observations and dynamics, we also face a parallel challenge in exploring these topics. It is evident that the scarcity of literature on these issues stems from both their inherently interdisciplinary nature and the fragmented state of the very limited research that does address them. Nevertheless, in taking on the challenge of contributing to this field, we have decided to develop the following explorative research questions to expand the accounting literature:

- How do non-EU businesses adapt their performance measurement systems to align with European regulations, particularly regarding transparency and comparability?
- How do best practices in performance measurement and disclosure differ between EU and non-EU firms, and what factors drive these differences?
- How do micro-level (firm-level) accountability practices influence meso level (sector or supply chain) comparability, particularly when comparing EU-based and non-EU-based businesses?
- What are the main barriers to achieving transparent and comparable sustainability measurements within the EU, and how do these barriers extend to non-EU firms attempting alignment?

This exploration of the CSRD and ESRS (comparability perspectives) becomes even more complex with the emergence of the 'omnibus' proposal, which introduces additional regulatory dimensions and challenges for businesses. Given the complex interactions between regulatory innovation, business practices, and political-institutional proposals, policymakers have moved toward the "omnibus" perspective. The Omnibus Package is part of a broader effort to enhance the efficiency and proportionality of EU regulations, particularly in sustainable finance reporting, corporate sustainability due diligence, and taxonomy compliance. One of its central objectives is to prevent excessive reporting burdens on smaller companies that might be indirectly affected by new supply chain requirements. The intervention is expected to introduce revisions to key sustainability frameworks such as the

CSRD, CSDDD, and the EU Taxonomy, incorporating simplifications while ensuring that their original goals remain intact.

The EU "Competitiveness Compass," officially released on January 29, 2025 (see draft published by the European Commission, 2025), outlines the strategic direction for the Omnibus Package. A key focus of the intervention is the proportional implementation of timelines, ensuring that businesses have adequate time to comply with new regulatory requirements. At the same time, the initiative emphasizes that the alignment of the required data with investor needs to reduce unnecessary complexity. Another central aspect is the targeted focus on high-impact activities, preventing the indiscriminate application of obligations that could disproportionately affect certain sectors. Additionally, Omnibus will introduce a new category of "small mid-caps," positioned between SMEs and large corporations. These companies (totaling approximately 31,000 in the EU) will benefit from tailored regulatory simplifications that acknowledge their specific challenges and resource constraints. While further details will be revealed with the official publication of the Competitiveness Compass, the planned changes indicate a substantial shift in the regulatory landscape rather than minor efficiency adjustments. The overall direction suggests a strong commitment to balancing regulatory rigor with business competitiveness, ensuring that sustainability and financial transparency requirements remain feasible while still contributing to the EU's broader economic and environmental objectives.

This latest innovation, or the expected modification of the current status quo, further increases the uncertainty and complexity of studying the dynamics of business competitiveness and, consequently, their performance within the European context. On the one hand, Europe stands at the forefront of sustainable production processes. On the other hand, it faces strong competitive pressure from other regions (see Sect. 3.2). To ensure that European businesses are not left disadvantaged in the market, regulators need to recalibrate their efforts. A similar perspective was also reflected in the recent Draghi Report on the topic, particularly concerning the effects of implementing decarbonization strategies without compromising the competitiveness of production systems (European Commission, 2024, p. 41).

By deferring reporting waves, raising CSRD thresholds, and introducing a value chain cap, the Omnibus Package shifts its implementation from a wide, rapid rollout to a staged, capacity-sensitive diffusion. Evidence from the NFRD shows that the effects of EU reporting mandates have been heterogeneous across Member States, with comparability gains depending on local administrative capacity and enforcement. A staged approach can therefore ameliorate or exacerbate cross-country gaps depending on supervisory coordination (Breijer & Orij, 2022). Economic consequences research similarly predicts that benefits will appear alongside adjustment and distributional costs, implying asynchronous convergence in the presence of divergent national capacities (Christensen et al., 2021). Overviews of the CSRD/ESRS architecture also stress that policy sequencing will be decisive in maintaining interoperability and comparability during simplification (Hummel, 2024). Following this reasoning, scholars could pursue questions such as

- How do deferrals, higher thresholds, and the value chain cap alter the timing and depth of accountability diffusion across Member States with different administrative capacities?
- What institutional complementarities or frictions arise between EU-level simplification and national transposition/enforcement, and how do these affect cross-border comparability?
- To what extent does the Omnibus pacing support competitiveness-consistent decarbonization, and how can this balance be operationalized in sectoral panels?

Deferring sector-specific ESRS and re-grading 300+ data points rebalances the demand from broad coverage to sequenced depth, with implications for assurance capacity and consistency. Prior work has documented jurisdictional competition and practice diversity in sustainability assurance, indicating that without supervisory alignment of evidence quality and materiality documentation, simplification risks weakening the information signal (Channuntapipat et al., 2020; Simnett et al., 2009). Recent reviews similarly warn that rapid growth in engagements has been met with uneven expertise and process variation, underscoring the need for EU-level coordination while sector rules are deferred (Hazaea et al., 2022). Following this reasoning, we propose the following research questions:

- How will deferring sector-specific ESRS shape sectoral transformation trajectories and supply/demand for limited vs. reasonable assurance across Member States?
- Which coordination mechanisms between standard setting and supervision best prevent data point re-grading from reducing the information signal on systemic risks?
- How do Omnibus choices on assurance trajectories interact with national assurance markets and costs passed to SMEs and "small mid-caps"?
- How will the propensity for the reasonable assurance option be impacted, given that the Omnibus Package requires limited assurance on sustainability reports? Moreover, how will this downside of the required assurance effort impact the quality of sustainability reports? Are these effects expected to be more pronounced in certain sectors that are more under stakeholder scrutiny?

The value chain cap aims to curb indiscriminate trickle-down requests while preserving the visibility of salient risks. Accounting scholarship shows that materiality remains contested and method-dependent, with documentation quality and evidence governance driving decision usefulness more than raw data point volume (Oll et al., 2025). Early evidence on double-materiality practice points to heterogeneous methods and depth, which Omnibus-enabled sequencing could amplify unless guided by robust process expectations (Dragomir et al., 2025). Policy analysis around the CSRD likewise supports phased hotspot coverage and shared infrastructures to overcome deep-tier bottlenecks. Several research questions arise from this reasoning:

- Which institutional incentives (guidance, tagging taxonomies, supervisory feedback) most effectively reduce deep-tier data bottlenecks while keeping proportionality?
- How does the value chain cap change information flows, and what metrics of "proportionality with integrity" can monitor whether it reduces duplication without degrading risk visibility?
- What is the EU-wide effect of VSME diffusion on data quality and comparability across sectors/regions?

Omnibus Package adjustments that interact with the EU Taxonomy (and banks' Green Asset Ratio) re-weight disclosure toward financial-sector usability, but simplified templates or exclusions can reduce the signal-to-noise ratio if underlying data points thin out. Research shows that taxonomy-related metrics are highly sensitive to design choices and materially shape investors' and banks' decision-making; hence, simplification must be evaluated for its capital-allocation effects (Alessi et al., 2024). Following this reasoning, the proposed areas for further investigation are

- How do Omnibus-driven changes propagate into banks' taxonomy KPIs, the cost of capital for non-CSRD firms, and allocation to high-impact activities?
- Do exemptions/simplified templates produce observable shifts in issuer disclosure strategies and investor coverage, and are the effects symmetric across Member States?

Raising thresholds and encouraging VSME baselines aim to protect SMEs, yet prior evidence indicates that relief can unintentionally create accountability gaps if buyers replace blanket requests with under-scoped ones rather than risk-prioritized demands. Systematic reviews on SME disclosure determinants highlight that institutional pressure and capability building jointly drive reporting quality (Lisi et al., 2023; O'Reilly et al., 2025). Therefore, to avoid two-tier transparency, the Omnibus equilibrium will hinge on collective enablement mechanisms (training, shared platforms) rather than relief alone. Following this reasoning and prior research on the transparency of multi-tier supply chains (Marttinen et al., 2022; Mena et al., 2013; Sauer & Seuring, 2019), we propose the following research questions:

- How do raised thresholds and the treatment of "small mid-caps" redistribute compliance costs and accountability along EU value chains in concentrated vs. fragmented sectors?
- Which public or collective capability-building arrangements most effectively convert Omnibus proportionality into SME enablement rather than burden shifting?

Narrowing the scope and staging of the sector rules will change how ESRS expectations travel to non-EU suppliers. Comparative accounting analyses emphasize that interoperability is most valuable when firms face multi-regime demands; trimming EU data points may ease short-run reconciliation with ISSB/GRI while also weakening cross-regime comparability if critical impact indicators become voluntary during phase-in (Hummel, 2024). In this vein, scholars could pursue the following research questions:

- How do Omnibus-revised ESRS (the new draft published in July 2025) demand transmission to non-EU tiers, and what patterns of selective alignment emerge across sectors with different governance types?
- Which coordination mechanisms with ISSB/GRI preserve comparability when EU re-grading alters the relative salience of disclosures?

The Omnibus Package is a competitiveness-oriented recalibration: Simplification can lower near-term compliance frictions, yet research cautions that oversimplification may blunt market discipline and dampen the real effects of mandates (Christensen et al., 2021). NFRD experience indicates that without adequate supervision and capacity, reforms can yield uneven improvements in disclosure quality and comparability (Breijer & Orij, 2022), making implementation quality the critical margin on which the Omnibus Package effect could be judged. This reasoning leads to the following research questions:

- Under what conditions does the Omnibus Package support competitiveness-consistent decarbonization, and how can we empirically distinguish burden relief from accountability dilution?
- How do Member-State industrial and procurement policies interact with Omnibus-calibrated reporting to generate learning effects or lock-ins, and with what cross-border spillovers?

With these issues and their interactions continuing to be impactful, there is clearly a need for social and environmental accounting scholars to deepen their understanding of both micro- and meso-level dynamics. Moreover, these perspectives, linked to the Draghi Report and Omnibus Package measures, can ultimately provide a foundation for further developing and expanding the framework presented in Fig. 5.1. The conclusion section offers some final considerations and a summary of the book's topics, themes, and vision.

References

Abbasi, M., & Nilsson, F. (2012). Themes and challenges in making supply chains environmentally sustainable. *Supply Chain Management: An International Journal, 17*(5), 517–530.

Adams, C. A. (2004). The ethical, social and environmental reporting-performance portrayal gap. *Accounting, Auditing & Accountability Journal, 17*(5), 731–757.

Adams, C. A. (2017). Conceptualising the contemporary corporate value creation process. *Accounting, Auditing & Accountability Journal, 30*(4), 906–931.

Adams, C. A., & McNicholas, P. (2007). Making a difference: Sustainability reporting, accountability and organisational change. *Accounting, Auditing & Accountability Journal, 20*(3), 382–402.

Ahi, P., & Searcy, C. (2013). A comparative literature analysis of definitions for green and sustainable supply chain management. *Journal of Cleaner Production, 52*, 329–341.

Ahmadov, T., Durst, S., Gerstlberger, W., & Kraut, E. (2023). SMEs on the way to a circular economy: Insights from a multi-perspective review. *Management Review Quarterly*, 1–34.

Alessi, L., Cojoianu, T., Hoepner, A. G., & Michelon, G. (2024). Accounting for the EU green taxonomy: Exploring its concept, data and analytics. *Accounting Forum, 48*(3), 365–373.

Appannan, J. S., Mohd Said, R., Ong, T. S., & Senik, R. (2023). Promoting sustainable development through strategies, environmental management accounting and environmental performance. *Business Strategy and the Environment, 32*(4), 1914–1930.

Arkoh, P., Costantini, A., & Scarpa, F. (2024). Determinants of sustainability reporting: A systematic literature review. *Corporate Social Responsibility and Environmental Management, 31*(3), 1578–1597.

Bai, C., Zhu, Q., & Sarkis, J. (2024). Circular economy and circularity supplier selection: A fuzzy group decision approach. *International Journal of Production Research, 62*(7), 2307–2330.

Bales, K., & Sovacool, B. K. (2021). From forests to factories: How modern slavery deepens the crisis of climate change. *Energy Research & Social Science, 77*, 102096. https://doi.org/10.1016/j.erss.2021.102096

Bayne, L. (2022). Understanding reporting boundaries in annual reports: A conceptual framework. *Accounting, Auditing & Accountability Journal, 35*(5), 1316–1348.

Beske, P., & Seuring, S. (2014). Putting sustainability into supply chain management. *Supply Chain Management: An International Journal, 19*(3), 322–331.

Beske-Janssen, P., Johnsen, T., Constant, F., & Wieland, A. (2023). New competences enhancing procurement's contribution to innovation and sustainability. *Journal of Purchasing and Supply Management, 29*(3), 100847.

Breijer, R., & Orij, R. (2022). The comparability of non-financial information: An exploration of the impact of the non-financial reporting directive (2014/95/EU). *Accounting in Europe, 19*(2), 332–361.

Brown, J., & Dillard, J. (2014). Integrated reporting: On the need for broadening out and opening up. *Accounting, Auditing & Accountability Journal, 27*(7), 1120–1156.

Brown, D., Boyd, D. S., Brickell, K., Ives, C. D., Natarajan, N., & Parsons, L. (2021). Modern slavery, environmental degradation, and climate change: Fisheries, field, forests and factories. *Environment and Planning. E, Nature and Space (Print), 4*(2), 191–207. https://doi.org/10.1177/2514848619887156

Busch, T., Barnett, M. L., Burritt, R. L., Cashore, B. W., Freeman, R. E., Henriques, I., York, J., et al. (2024). Moving beyond "the" business case: How to make corporate sustainability work. *Business Strategy and the Environment, 33*(2), 776–787.

Caccialanza, A., & Marinoni, M. (2023). The spread of sustainability reporting in the Italian food manufacturing context. In *Sustainable Transition of Meat and Cured Meat Supply Chain: A Transdisciplinary Approach* (pp. 243–264). Cham: Springer Nature Switzerland.

Caccialanza, A., & Torelli, R. (2024). The role of trade associations in promoting corporate sustainability transition and reporting: A case study in the food supply chain. *Corporate Social Responsibility and Environmental Management.*

Camilleri, M. A. (2020). European environment policy for the circular economy: Implications for business and industry stakeholders. *Sustainable Development, 28*(6), 1804–1812.

Channuntapipat, C., Samsonova-Taddei, A., & Turley, S. (2020). Variation in sustainability assurance practice: An analysis of jurisdictional competition. *The British Accounting Review, 52*(2), 100843.

Chief Sustainability Officers Network. (2024). *Post published by Philipp Schröder, Ph.D.* https://www.linkedin.com/feed/update/urn:li:activity:7284469627521253376?utm_source=share&utm_medium=member_desktop.

Cho, C. H., & Patten, D. Ms. (2007). The role of environmental disclosures as tools of legitimacy: A research note. *Accounting, Organizations and Society, 32*(7–8), 639–647.

Christensen, H. B., Hail, L., & Leuz, C. (2021). Mandatory CSR and sustainability reporting: Economic analysis and literature review. *Review of Accounting Studies, 26*(3), 1176–1248.

Clarkson, P. M., Li, Y., Richardson, G. D., & Vasvari, F. P. (2008). Revisiting the relation between environmental performance and environmental disclosure: An empirical analysis. *Accounting, Organizations and Society, 33*(4–5), 303–327.

Daniel, S. J., Cieslewicz, J. K., & Pourjalali, H. (2012). The impact of national economic culture and country-level institutional environment on corporate governance practices: Theory and empirical evidence. *Management International Review, 52*, 365–394.

Das, D. (2017). Development and validation of a scale for measuring sustainable supply chain management practices and performance. *Journal of Cleaner Production, 164*, 1344–1362.

Delmas, M., & Toffel, M. W. (2004). Stakeholders and environmental management practices: An institutional framework. *Business Strategy and the Environment, 13*(4), 209–222.

Dragomir, V.-D., Dumitru, M., Chersan, I.-C., Gorgan, C., & Păunescu, M. (2025). Double materiality disclosure as an emerging practice: The assessment process, impacts, risks, and opportunities. *Accounting in Europe, 22*(1), 103–140.

Dutta, P., & Dutta, A. (2024). Does corporate environmental performance affect corporate biodiversity reporting decision? The Finnish evidence. *Journal of Applied Accounting Research, 25*(1), 24–41.

EFRAG. (2022). *[Draft] EFRAG IG 3 list of ESRS data points (last update: 23/12/2022)*. https://efrag.sharefile.com/share/view/s1a12c193b86d406e90b1bcd7b6bb8f6f/fo37c90b-9d9b-4432-a76b-27760cfcc01b

EFRAG. (2024a). *EFRAG IG 1 materiality assessment (last update: May 2024)*. https://www.efrag.org/sites/default/files/sites/webpublishing/SiteAssets/IG%201%20Materiality%20Assessment_final.pdf

EFRAG. (2024b). *Voluntary SME standard (VSME)*. European Financial Reporting Advisory Group (EFRAG). https://www.efrag.org/sites/default/files/sites/webpublishing/SiteAssets/VSME%20Standard.pdf.

El-Garaihy, W. H., Farag, T., Al Shehri, K., Centobelli, P., & Cerchione, R. (2024). Driving sustainability in supply chain management for a more inclusive and responsible future. *International Journal of Productivity and Performance Management, 73*(1), 43–84.

European Commission. (2024). *EU competitiveness: Looking ahead—Report by Mario Draghi*. https://commission.europa.eu/topics/strengthening-european-competitiveness/eu-competitiveness-looking-ahead_en

European Commission. (2025). Communication from the commission to the European Parliament, the Council, the European Economic and Social Committee and the committee of the regions: A competitiveness compass for the EU. https://eur-lex.europa.eu/legal-content/EN/TXT/?uri=celex:52024DC0077

Galli, D., Torelli, R., & Caccialanza, A. (2024). Sustainability performance and sustainability reporting in SMEs: A love affair or a fight? *Journal of Management & Organization, 30*(3), 574–599.

Garcia-Torres, S., Rey-Garcia, M., & Sáenz, J. (2024). Enhancing sustainable supply chains through traceability, transparency and stakeholder collaboration: A quantitative analysis. *Business Strategy and the Environment, 33*(7), 7607–7629.

Global Reporting Initiative. (2023, November 30). *GRI standards and ESRS draft interoperability index*. Global Reporting Initiative. https://www.globalreporting.org/media/z2vmxbks/gri-standards-and-esrs-draft-interoperability-index_20231130-final.pdf.

Gold, S., & Heikkurinen, P. (2018). Transparency fallacy: Unintended consequences of stakeholder claims on responsibility in supply chains. *Accounting, Auditing & Accountability Journal, 31*(1), 318–337.

Govindan, K., & Bouzon, M. (2018). From a literature review to a multi-perspective framework for reverse logistics barriers and drivers. *Journal of Cleaner Production, 187*, 318–337.

Govindan, K., Shaw, M., & Majumdar, A. (2021). Social sustainability tensions in multi-tier supply chain: A systematic literature review towards conceptual framework development. *Journal of Cleaner Production, 279*, 123075.

Grabs, J., Carodenuto, S., Jespersen, K., Adams, M. A., Camacho, M. A., Celi, G., Stone, E., et al. (2024). The role of midstream actors in advancing the sustainability of Agri-food supply chains. *Nature Sustainability, 7*(5), 527–535.
Guidi, M., Vitali, S., Giuliani, M., & Chiucchi, M. S. (2024). Understanding the interplay between sustainability strategy and the approach to sustainability reporting in SMEs. *Journal of Management & Organization*, 1–20.
Han, D., Kalantari, M., & Rajabifard, A. (2024). The development of an integrated BIM-based visual demolition waste management planning system for sustainability-oriented decision-making. *Journal of Environmental Management, 351*, 119856.
Hazaea, S. A., Zhu, J., Khatib, S. F. A., Bazhair, A. H., & Elamer, A. A. (2022). Sustainability assurance practices: A systematic review and future research agenda. *Environmental Science and Pollution Research, 29*(4), 4843–4864.
Hummel, K. (2024). An overview of corporate sustainability reporting in the EU: CSRD/ESRS and global standard-setting. *Accounting in Europe, 21*(3), 320–355.
Kalogirou, V., Stasis, A., & Charalabidis, Y. (2022). Assessing and improving the national interoperability frameworks of European Union member states: The case of Greece. *Government Information Quarterly, 39*(3), 101716.
Kazancoglu, Y., Ozkan-Ozen, Y. D., Sagnak, M., Kazancoglu, I., & Dora, M. (2023). Framework for a sustainable supply chain to overcome risks in transition to a circular economy through industry 4.0. *Production Planning & Control, 34*(10), 902–917.
Khalid, R. U., Seuring, S., Beske, P., Land, A., Yawar, S. A., & Wagner, R. (2015). Putting sustainable supply chain management into base of the pyramid research. *Supply Chain Management: An International Journal, 20*(6), 681–696.
Koberg, E., & Longoni, A. (2019). A systematic review of sustainable supply chain management in global supply chains. *Journal of Cleaner Production, 207*, 1084–1098.
Lai, A., & Stacchezzini, R. (2021). Organisational and professional challenges amid the evolution of sustainability reporting: a theoretical framework and an agenda for future research. *Meditari Accountancy Research, 29*(3), 405–429.
Leite, E., Koporcic, N., & Markovic, S. (2024). Corporate sustainability reporting: Shifting from optional due diligence to mandatory duty. *Business Ethics, the Environment & Responsibility.*
Lisi, S., Mignacca, B., & Grimaldi, M. (2023). Non-financial reporting and SMEs: A systematic review, research agenda, and novel conceptualization. *Journal of Management & Organization*, 1–23.
Luo, L., Tang, Q., & Lan, Y. C. (2013). Comparison of propensity for carbon disclosure between developing and developed countries: A resource constraint perspective. *Accounting Research Journal, 26*(1), 6–34.
Marttinen, K., Kauppi, K., & Kuula, M. (2022). Fostering firms' ability to cascade sustainability through multi-tier supply chains. *International Journal of Operations & Production Management, 42*(13), 151–180.
Mena, C., Humphries, A., & Choi, T. Y. (2013). Toward a theory of multi-tier supply chain management. *Journal of Supply Chain Management, 49*(2), 58–77.
Moggi, S. (2019). Social and environmental reports at universities: A Habermasian view on their evolution. *Accounting Forum, 43*(3), 283–326.
Moggi, S., Pierce, P., & Bernardi, N. (2022). From sustainability to thrivability: A novel framework for entrepreneurial ecosystems. *International Entrepreneurship and Management Journal, 18*(2), 829–853.
Neri, A., Negri, M., Cagno, E., Franzò, S., Kumar, V., Lampertico, T., & Bassani, C. A. (2023). The role of digital technologies in supporting the implementation of circular economy practices by industrial small and medium enterprises. *Business Strategy and the Environment, 32*(7), 4693–4718.
Oll, J., Spandel, T., Schiemann, F., & Akkermann, J. (2025). The concept of materiality in sustainability reporting: From essential contestation to research opportunities. *Sustainability Accounting, Management and Policy Journal, 16*(2), 321–350.

O'Reilly, S., Mac An Bhaird, C., Gorman, L., & Brennan, N. M. (2025). Accounting practitioners' perspectives on small-and medium-sized enterprises' environmental sustainability reporting. *Journal of Applied Accounting Research, 26*(6), 26–46.

Pagell, M., & Shevchenko, A. (2014). Why research in sustainable supply chain management should have no future. *Journal of Supply Chain Management, 50*(1), 44–55.

Pickering, J., Bäckstrand, K., & Schlosberg, D. (2020). Between environmental and ecological democracy: Theory and practice at the democracy-environment nexus. *Journal of Environmental Policy & Planning, 22*(1), 1–15.

PwC. (2024). *PwC's Global CSRD Survey 2024 - The promise and reality of CSRD reporting.* https://www.pwc.com/gx/en/issues/esg/global-csrd-survey.html?mod=djemSustainableBusinessPro

Rajeev, A., Pati, R. K., Padhi, S. S., & Govindan, K. (2017). Evolution of sustainability in supply chain management: A literature review. *Journal of Cleaner Production, 162*, 299–314.

Sauer, P. C., & Seuring, S. (2019). Extending the reach of multi-tier sustainable supply chain management–insights from mineral supply chains. *International Journal of Production Economics, 217*, 31–43.

Scagnelli, S. D., Corazza, L., & Cisi, M. (2013). How SMEs disclose their sustainability performance. Which variables influence the choice of reporting guidelines? In *Accounting and control for sustainability* (pp. 77–114). Emerald Group Publishing Limited.

Simnett, R., Vanstraelen, A., & Chua, W. F. (2009). Assurance on sustainability reports: An international comparison. *The Accounting Review, 84*(3), 937–967.

Trianni, A., Cagno, E., & Neri, A. (2017). Modelling barriers to the adoption of industrial sustainability measures. *Journal of Cleaner Production, 168*, 1482–1504. https://doi.org/10.1016/j.jclepro.2017.07.244

Troisi, R., De Simone, S., & Franco, M. (2024). Illegal firm behaviour and environmental hazard: The case of waste disposal. *European Management Review, 21*(3), 605–617.

Turner, L., & Weirich, T. R. (2023). Expanding the concept of materiality to environmental, social, and governance: Audit issues and implications. *Current Issues in Auditing, 17*(1), A50–A58.

Vegter, D., van Hillegersberg, J., & Olthaar, M. (2023). Performance measurement system for circular supply chain management. *Sustainable Production and Consumption, 36*, 171–183.

Wolf, J. (2014). The relationship between sustainable supply chain management, stakeholder pressure and corporate sustainability performance. *Journal of Business Ethics, 119*, 317–328.

Chapter 6
Conclusions

6.1 Research Question and Synthesis of Findings

This monograph asks how the European sustainability reporting architecture—centered on the CSRD/ESRS and recalibrated by the Omnibus simplification package—reconfigures accountability from the single firm (micro) to the supply chain (meso). Likewise, it considered the institutional and data mechanisms through which this accountability can be made proportionate, comparable, and decision-useful. The analysis proceeded across the literature reviewed in the book and is anchored in the theoretical framework developed in Chap. 5, which integrates (i) the sustainability of the value chain, (ii) the accountability of the value chain, and (iii) the intervening role of the institutional level (EFRAG/ESRS) into a single meso-level lens (Fig. 5.1 in Chap. 5). The framework clarifies how disclosure standards and transparency obligations can translate into supply chain governance routines: how the "arrows" between sustainability practices, accountability processes, and institutional levers coordinate expectations and evidence along upstream and downstream tiers according to the several issues involved in those three macro-dimensions.

Three findings emerge. First, the ESRS formalizes meso-level accountability by blending a value chain lens with double materiality (EFRAG, 2022). In this case, the impacts, risks, and opportunities (IROs) are to be identified "where they actually occur" across tiers, rather than only inside the reporting (corporate) boundaries (EFRAG, 2024a). Early practice confirms both momentum and strain: Value chain analysis is the least mature area of implementation; most firms are still refining mapping granularity, and many start at tier-1 with only a gradual extension to deeper tiers (EFRAG, 2024, p. 6). In parallel, many organizations are beginning to embed evidence-based double materiality into decision processes, yet data integration across the chain remains the binding constraint on comparability (EFRAG, 2024b). These observations substantiate the framework's core claim: accountability

A. Caccialanza, *Sustainability and Accountability of Supply Chains*, SIDREA Series in Accounting and Business Administration,
https://doi.org/10.1007/978-3-032-16879-5_6

can only be scaled to the meso level insofar as chain-specific evidence is produced and governed (EFRAG, 2024; EFRAG, 2025).

Second, the Omnibus Package strategically re-sequences and right-sizes implementations without abandoning the core. It (a) defers obligations for later waves, (b) raises CSRD thresholds (limiting scope to firms with higher thresholds in terms of employees), (c) introduces a value chain cap to contain indiscriminate "trickle-down" data requests, (d) narrows due diligence to direct partners, (e) defers sector-specific ESRS to 30 June 2026, and (f) re-grades 300+ data points from "shall disclose" to "voluntary/subject-to-materiality" or guidance. All of this is being done to explicitly streamline the first cycle and align demands with current data capabilities. The package operates alongside the VSME voluntary standard for non-listed SMEs, using a shared baseline to stabilize information requests along chains and reduce duplication, especially for suppliers outside the CSRD perimeter (Chap. 3; §4.4). This recalibration does not dilute the accountability core—double materiality remains the backbone—but changes where and how quickly comparability can be credibly achieved.

Third, the diffusion of accountability is bound by data availability and quality across tiers. The supply chain literature has long shown that salient environmental and social risks often sit beyond tier-1, where visibility, incentives, and measurement capacity are weakest; digital traceability and collaboration can help, but governance choices—who collects what, at which tier, with which assurance—determine whether disclosure becomes decision-useful (and comparable) or remains symbolic (e.g., transparency fallacy). The empirical and review work cited here confirms this point: Institutional pressures do shape practice (Kauppi & Hannibal, 2017; Wu et al., 2013; Zhu & Sarkis, 2007), but comparability depends on evidence governance, not on data point counts alone (Garcia-Torres et al., 2024; Gold & Heikkurinen, 2018).

Together, these findings specify how the ESRS are more than a standard or rulebook: They constitute, in the medium to long term, a meso-level governance architecture. The novelty of this monograph is precisely in tracing that architecture into the chain, showing how institutional design (i.e., ESRS, Omnibus Package, VSME) can translate stakeholder and regulatory expectations into chain-aware accountability routines, provided that materiality is executed with chain-specific evidence and proportionality tools are used to prioritize risk rather than lower ambition.

6.2 Contributions

6.2.1 Theoretical Implications

The theoretical contribution lies in reframing double materiality as a method for meso-level accountability, rather than as a label for additional disclosure. In the framework of Chap. 5, the sustainability of the value chain and accountability of the value chain are linked by IRO-first identification grounded in place-based evidence that can travel across tiers. The institutional level (EFRAG/ESRS) specifies data points, boundaries, and assurance trajectories that convert that evidence into comparable, decision-useful information (i.e., EFRAG IG-1 on materiality; IG-3 on data points). This moves the debate beyond the generic "does sustainability pay?" question toward a theory of how accountability is allocated and evidenced across interfirm relationships (and how it becomes comparable). In this sense, the monograph stimulates the debate with the proposed framework's mechanics.

A second theoretical contribution is the institutional reading of proportionality. The Omnibus levers—higher thresholds, the value chain cap, a slower sectoral ramp-up, and data point re-grading—are interpreted here as design choices that modulate information cascades while keeping the normative core intact. This perspective, combined with evidence of early practice (EFRAG, 2024), shows why the pace and focus of accountability diffusion can be tuned without collapsing ESRS's intent. It also clarifies the expected externalities for non-EU suppliers trading into EU chains—a theme raised in the original conclusions (Kazancoglu et al., 2021; Tsai et al., 2021). By situating proportionality inside a meso-level framework, this study contributes a governance-of-disclosure account rather than a purely compliance-based one.

Finally, the contribution to the data-for-sustainability debate is to connect comparability to evidence governance. The review shows that institutional forces trigger adoption (Imtiaz Ferdous et al., 2019; Martinez-Ferrero & García-Sánchez, 2017; Modell, 2012), but decision usefulness depends on the credibility and coverage of chain-level indicators, especially beyond tier-1. The synthesis of sectoral and cross-sectoral work (Garcia-Torres et al., 2024; Gold & Heikkurinen, 2018) underscores where standardization is feasible and deep-tier measurement remains a research and practice frontier, as mapped explicitly in Chap. 5's agenda.

6.2.2 Managerial Implications

For managers, this book interprets ESRS adoption as chain-aware management, not mere compliance. In the vocabulary of Chap. 5, implementing IRO-first materiality means (i) mapping salient hotspots beyond tier-1 where feasible, (ii) documenting evidence-based significance, and (iii) linking each IRO to policies, actions, targets, and metrics so that disclosure mirrors real decisions (on this point, also see the EFRAG Implementation evidence). Used this way, ESRS reporting becomes a control instrument that aligns procurement, risk, and strategy across tiers rather than an after-the-fact communication exercise.

Second, the Omnibus + VSME combination should be operationalized as a proportional data architecture. The VSME provides a baseline that stabilizes supplier requests. The value chain cap operationalizes risk-based prioritization rather than blanket questionnaires, thereby reducing duplication while preserving the accountability signal (Chaps. 3 and 4; VSME; Omnibus). This directly addresses a key tension in the literature between transparency and burden shifting along the chain (Gold & Heikkurinen, 2018) and speaks to the original conclusions' emphasis on inbound/outbound externalities (Kazancoglu et al., 2021; Tsai et al., 2021).

Third, managers should consider data readiness as a strategic capability. Early EU evidence points to data retrieval hurdles and reliance on phase-ins; investing in traceability, supplier enablement, and assurance-ready data models will determine whether firms can deliver ESRS comparability and respond credibly to stakeholder scrutiny (EFRAG, 2024). This speaks directly to concerns in the supply chain literature about the quality of information and the risks of symbolic disclosure absent evidence of governance (Garcia-Torres et al., 2024; Wu et al., 2013; Zhu & Sarkis, 2007).

6.3 Limitations and Concluding Remarks

This contribution is conceptual and oriented toward the systematization of the status quo in the debate. Parts of the institutional analysis draw on exposure drafts and evolving guidance (amended ESRS 2, July 2025); particular details may change as the Omnibus process advances. The focus is EU-centered and does not supply cross-jurisdictional datasets. Moreover, because ESRS-compliant annual reports were scarce at project inception, we cannot offer a longitudinal assessment of reporting quality or outcomes. While initial practice signals help set testable hypotheses, they do not substitute for ex-post evaluation (amended ESRS 2 ED; EFRAG, 2024; Martínez-Ferrero & García-Sánchez, 2017; Modell, 2012). Sustainable supply chain management can promote the spread of a sustainability-oriented culture not only within the focal company but also to its external stakeholders (e.g., suppliers and consumers). In line with this trending topic, the present work contributes to the literature by synthesizing the institutional (and largely external) drivers of sustainability performance disclosure within the supply chain management debate (Imtiaz Ferdous et al., 2019; Kauppi & Hannibal, 2017; Wu et al., 2013; Zhu & Sarkis, 2007). This entailed reviewing previous studies on sustainable supply chain management and discussing existing drivers and barriers, considering the draft ESRS. By requiring firms to engage in thorough reporting of supply chain responsibility, the EU seeks to encourage businesses to embrace sustainable and ethical operational approaches that foster long-term societal and environmental welfare. Additionally, the double materiality framework adopted in the preliminary ESRS underscores the EU's leadership position in advancing worldwide sustainability norms and establishing criteria for corporate responsibility (Dodsworth et al., 2023; Setia et al., 2024). As a result, the latest draft ESRS represents a substantial attempt to bolster supply chain accountability within the European context, offering valuable resources to both researchers and industry practitioners.

This monograph's novelty lies in centering on the sustainability and accountability of supply chains, which is not yet a diffused perspective among accounting firms. To this end, we show how ESRS (despite the intervention of the Omnibus

Package) creates a governance architecture that can transmit accountability along upstream and downstream relationships. This monograph's main contribution is in stimulating discussions about the advantages and concerns of the ESRS approach at a European scale, particularly regarding the inbound and outbound external impacts (Kazancoglu et al., 2021; Tsai et al., 2021). Specifically, it proposes a framework that encompasses the role of EFRAG and ESRS in assessing said impacts on both EU and non-EU businesses, as well as outlines the expected effects on sustainable supply chain managerial practices. It restates the research question in operational terms: what institutional levers and evidence practices make meso-level accountability work? It then answers with a framework that connects supply chain-specific materiality to proportional, standardized data points, showing how these design choices can produce comparable, decision-useful information without overwhelming SMEs. The synthesis of institutional, accounting, and SSCM literature explains why institutional pressures drive adoption (Imtiaz Ferdous et al., 2019; Kauppi & Hannibal, 2017; Wu et al., 2013; Zhu & Sarkis, 2007), how double materiality stabilizes the conceptual core (Dodsworth et al., 2023; Setia et al., 2024), and where accountability diffusion is most fragile, precisely where data are hardest to obtain beyond tier-1 (Garcia-Torres et al., 2024; Gold & Heikkurinen, 2018). The contribution theorizes a meso-level accountability mechanism and positions the work within and against the literature it mobilizes, setting an empirical agenda for pre/post-ESRS comparisons and deep-tier evidence building in European and non-European chains (Kazancoglu et al., 2021; Tsai et al., 2021).

Further research should expand on the proposed themes by broadening the involved actor categories (e.g., consumers or other standard setters) or discussing these actors' feedback on the published drafts. Alternatively, scholars could pursue case studies—whether involving individual enterprises affected by the new regulations or a collection of cases that illustrate the supply chain effects—to further enrich the emerging literature on EFRAG's ESRS. Because the present research did not have access to financial statements prepared in accordance with the ESRS during the preliminary stage of the work (the fall of 2024), we could not evaluate the outcomes of companies' sustainability performance or the quality of their reporting. Future studies may gain deeper insights by comparing pre- and post-ESRS reports.

In terms of theory, this study integrates the literature on supply chain management and sustainability reporting, answering a call for more interdisciplinary research in accounting studies (Garcia-Torres et al., 2024; Gold & Heikkurinen, 2018). On a practical level, this work stresses the interdependence among economic actors involved in a production chain, in the hopes of encouraging practitioners to understand the logic and value of sustainability reporting practices. Both theoretical and practical debates are important in assessing the externalities of European legislation. Ideally, this research will help shift the paradigm about sustainability performance accountability from individual firm performance (micro-level) to an integrated assessment of performances achieved across the supply chain (meso-level).

References

Adams, C. A. (2004). The ethical, social, and environmental reporting-performance portrayal gap. *Accounting, Auditing & Accountability Journal, 17*(5), 731–757.

Dodsworth, C., Bourne, E., Schlich, B. L., & Kooroshy, J. (2023). Financial, double, or dynamic? Theories of ESG materiality and practitioner approaches. *Journal of Impact & ESG Investing, 4*(2).

EFRAG. (2022). *[Draft] EFRAG IG 3 list of ESRS data points* (last update: 23/12/2022). https://efrag.sharefile.com/share/view/s1a12c193b86d406e90b1bcd7b6bb8f6f/fo37c90b-9d9b-4432-a76b-27760cfcc01b

EFRAG. (2024). *Implementation of ESRS: Initial observed practices from selected companies.* https://www.efrag.org/sites/default/files/media/document/2024-07/EFRAG_ESRS%20initial%20observed%20practices%20Q2%202024%20final%20version.pdf

EFRAG. (2024a). *EFRAG IG 1 materiality assessment* (last update: May 2024). https://www.efrag.org/sites/default/files/sites/webpublishing/SiteAssets/IG%201%20Materiality%20Assessment_final.pdf

EFRAG. (2024b). *Voluntary SME standard (VSME).* European Financial Reporting Advisory Group (EFRAG). https://www.efrag.org/sites/default/files/sites/webpublishing/SiteAssets/VSME%20Standard.pdf

EFRAG. (2025). *[draft] amended ESRS 2—Exposure draft—July 2025.* European Financial Reporting Advisory Group. https://www.efrag.org/sites/default/files/media/document/2025-07/Amended_ESRS_Exposure_Draft_July_2025_ESRS_2.pdf

Garcia-Torres, S., Rey-Garcia, M., & Sáenz, J. (2024). Enhancing sustainable supply chains through traceability, transparency and stakeholder collaboration: A quantitative analysis. *Business Strategy and the Environment, 33*(7), 7607–7629.

Gold, S., & Heikkurinen, P. (2018). Transparency fallacy: Unintended consequences of stakeholder claims on responsibility in supply chains. *Accounting, Auditing & Accountability Journal, 31*(1), 318–337.

Imtiaz Ferdous, M., Adams, C. A., & Boyce, G. (2019). Institutional drivers of environmental management accounting adoption in public sector water organisations. *Accounting, Auditing & Accountability Journal, 32*(4), 984–1012.

Kauppi, K., & Hannibal, C. (2017). Institutional pressures and sustainability assessment in supply chains. *Supply Chain Management: An International Journal, 22*(5), 458–472.

Kazancoglu, I., Sagnak, M., Kumar Mangla, S., & Kazancoglu, Y. (2021). Circular economy and the policy: A framework for improving the corporate environmental management in supply chains. *Business Strategy and the Environment, 30*(1), 590–608.

Martínez-Ferrero, J., & García-Sánchez, I. M. (2017). Coercive, normative and mimetic isomorphism as determinants of the voluntary assurance of sustainability reports. *International Business Review, 26*(1), 102–118.

Modell, S. (2012). Strategy, political regulation and management control in the public sector: Institutional and critical perspectives. *Management Accounting Research, 23*(4), 278–295.

Setia, N., Abhayawansa, S., Joshi, M., & Wasantha Pathiranage, N. (2024). Shifting perspectives: Unveiling the dual nature of sustainability materiality in integrated reports. *Meditari Accountancy Research*, (head-off-print).

Tsai, F. M., Bui, T. D., Tseng, M. L., Ali, M. H., Lim, M. K., & Chiu, A. S. (2021). Sustainable supply chain management trends in world regions: A data-driven analysis. *Resources, Conservation and Recycling, 167*, 105421.

Wu, T., Daniel, E. M., Hinton, M., & Quintas, P. (2013). Isomorphic mechanisms in manufacturing supply chains: A comparison of indigenous Chinese firms and foreign-owned MNCs. *Supply Chain Management: An International Journal, 18*(2), 161–177.

Zhu, Q., & Sarkis, J. (2007). The moderating effects of institutional pressures on emergent green supply chain practices and performance. *International Journal of Production Research, 45*(18–19), 4333–4355.